# FORTSCHRITTE DER PHYSIK

HERAUSGEGEBEN IM AUFTRAGE DER PHYSIKALISCHEN GESELLSCHAFT

DER

DEUTSCHEN DEMOKRATISCHEN REPUBLIK

VON F. KASCHLUHN, A. LÖSCHE, R. RITSCHL UND R. ROMPE

---

HEFT 5 · 1976 · BAND 24

AKADEMIE-VERLAG · BERLIN

EVP 10,— M

31728

BEZUGSMÖGLICHKEITEN

Bestellungen sind zu richten

- **in der DDR** an eine Buchhandlung oder an den **Akademie-Verlag**, DDR-108 Berlin, Leipziger Straße 3—4
- **im sozialistischen Ausland** an eine Buchhandlung für fremdsprachige Literatur oder an den zuständigen Postzeitungsvertrieb
- **in der BRD und Westberlin** an eine Buchhandlung oder an die Auslieferungsstelle KUNST UND WISSEN, Erich Bieber, 7 Stuttgart 1, Wilhelmstraße 4—6
- **in Österreich** an den Globus-Buchvertrieb, 1201 Wien, Höchstädtplatz 3
- **im übrigen Ausland** an den Internationalen Buch- und Zeitschriftenhandel; den Buchexport, Volkseigener Außenhandelsbetrieb der Deutschen Demokratischen Republik, DDR-701 Leipzig, Postfach 160, oder

  an den Akademie-Verlag, DDR-108 Berlin, Leipziger Straße 3—4

Zeitschrift „Fortschritte der Physik"

Herausgeber: Prof. Dr. Frank Kaschluhn, Prof. Dr. Artur Lösche, Prof. Dr. Rudolf Ritschl, Prof. Dr. Robert Rompe, im Auftrag der Physikalischen Gesellschaft der Deutschen Demokratischen Republik.
Verlag: Akademie-Verlag, DDR-108 Berlin, Leipziger Straße 3—4; Fernruf: 2200441; Telex-Nr. 114420; Postscheckkonto: Berlin 35021; Bank: Staatsbank der DDR, Berlin, Konto-Nr.: 6836-26-20712.
Chefredakteur: Dr. Lutz Rothkirch.
Anschrift der Redaktion: Sektion Physik der Humboldt-Universität zu Berlin, DDR-104 Berlin, Hessische Straße 2.
Veröffentlicht unter der Lizenznummer 1324 des Presseamtes beim Vorsitzenden des Ministerrates der Deutschen Demokratischen Republik.
Gesamtherstellung: VEB Druckhaus „Maxim Gorki", DDR-74 Altenburg, Carl-von-Ossietzky-Straße 30/31.
Erscheinungsweise: Die Zeitschrift „Fortschritte der Physik" erscheint monatlich. Die 12 Hefte eines Jahres bilden einen Band.
Bezugspreis je Band: 180,— M zuzüglich Versandspesen (Preis für die DDR: 120,— M). Preis je Heft 15,— M (Preis für die DDR: 10,— M).
Bestellnummer dieses Heftes: 1027/24/5.

Fortschritte der Physik 24, 259–323 (1976)

# Amplitude Reconstruction for Usual Quasi Two Body Reactions with Unpolarized or Polarized Target

M. G. Doncel* and P. Minnaert**

*CERN, Geneva*

L. Michel

*IHES, Bures-sur-Yvette*

**Abstract**

Tools for measuring joint polarization and polarization transfer are gathered. They allow the direct reconstruction of amplitudes in numerous quasi two body reactions with spinless beam and unpolarized or polarized target. Eight simple types of such reactions are worked out one by one; the practical results are summarized in Tables.

## 1. Introduction

Several experimental papers have been published [*1, 2, 3, 4*] in which quasi two body reaction amplitudes are reconstructed and tabulated. Old [*5, 6, 7*] and recent [*1, 8*] theoretical papers have described the method for reconstructing the amplitudes from the data in some simple cases. But each paper, limited to a particular reaction, introduces its own arbitrary conventions and this may obscure the future comparison of the experimental results for similar reactions. This paper aims to a systematics of amplitude reconstruction. It introduces a general terminology, it uses the most commonly accepted quantization conventions and it presents in details the practical method of amplitude reconstruction in the most usual reactions with unpolarized or polarized target.

We focus our attention on quasi two body reactions. We suppose therefore that resonances can be detected (we shall not enter into the problems of background separation) and that they have well known spins and parities. Furthermore, we restrict ourselves to a direct, model independent amplitude reconstruction, based only on angular momentum and parity conservation. Hence we shall not make use of other first principles as unitarity and analyticity, nor shall we relate amplitudes at different $s$ and $t$ values. Finally, this paper is limited to the simple initial state with a spinless meson beam and a spin 1/2 (polarized or unpolarized) baryon target.

* Departamento de Física Teórica — Universidad Autónoma de Barcelona, Bellaterra (Barcelona).
** Laboratoire de Physique Théorique — Université Bordeaux I — 33170 — GRADIGNAN.

In Table 1 we list the most common reaction types and tabulate their number of amplitudes and observables for different target polarizations. The reaction type refers uniquely to the spins of the final particles and to the possibility of analyzing their polarization from their decay. We assume that for the spin $l$ meson resonance only the even polarization can be measured (e.g. $\rho \to \pi\pi$, $K^* \to K\pi$), i.e. we do not consider the case of, e.g. $A_1 \to \rho\pi$. On the contrary, for the baryonic spin $j$ resonance we consider both cases: i) measurement of even polarization only ($j^e$), this means no polarization measurement in the case $1/2^e$ (e.g. nucleons), and analysis of the simple, parity conserving decay in the case $3/2^e$, (e.g. $\Delta \to N\pi$). ii) Measurement of the whole polarization ($j$) by analysis of the parity violating decay of spin 1/2 baryons (e.g. $\Lambda \to p\pi$) or by analysis of the cascade decay of spin 3/2 resonances (e.g. $\Sigma^* \to \Lambda\pi$, $\Lambda \to p\pi$).
The tabulated number of amplitudes is the number of *real* independent amplitudes disregarding an overall phase.
For the observables the number tabulated in columns $U$ and $T$ is the number of real and imaginary components of a priori non vanishing multipole parameters and polarization transfer multipole parameters (cf. sect. 2 and 3). As we shall see, at fixed energy and momentum transfer they are not independent, they satisfy some linear and non linear (rank) constraints[1]). For this reason the number of observables is presented as the sum of two numbers: the number of independent observables + the number of constraints (when the second number is zero it has been omitted). Column $U$ corresponds to an unpolarized target experiment, column $T$ corresponds to a transversally polarized target experiment; the observables of column $T$ includes of course the observables of column $U$. In column $L$ is shown the number of independent observables of constraints that must be added to column $U$ for an experiment with pure longitudinal polarization or to column $T$ for an experiment with transverse and longitudinal polarization. Of course the number of independent observables can never the bigger than the number of amplitudes, but often it is one unit smaller; this is the case when the polarization of one spinning initial or final particle is not at all considered. This is an application of a general theorem derived by Simonius [9], cf. Appendix 2. Practically the ghost amplitude appears as a relative phase between two sets of transversity amplitudes.
An analysis of the numbers in Table 1 suggests the following comments:

i) For higher spins the number of amplitudes increases linearly while the number of observables increases quadratically. When the number of observables becomes much bigger than the number of amplitudes it seems reasonable to communicate the amplitudes themselves, in so far as the statistics of the experiment allows their reconstruction.
ii) For some types of reactions all amplitudes but one can be reconstructed with unpolarized target, while for others amplitude reconstruction requires polarized target. For both reactions types, the most simplest cases are those for which the number of observables in columns $U$ and $T$ are set in boxes. It would be reasonable to give priority for the reconstruction of amplitudes to these types of reactions.
iii) As mentioned above, in all reaction types with unpolarized target and in those with a spin $1/2^e$ particle, it remains one ghost amplitude, the relative phase between two sets of transversity amplitudes. Since helicity amplitudes are linear combinations of amplitudes in both sets, they are ghosts too. Furthermore, parity conservation in the reaction has a simpler form in transversity quantization, and the observables are closer to the transversity amplitudes. All these arguments favor the reconstruction of transversity amplitudes. However to facilitate the comparison with models which use helicity amplitudes, for each reaction type we give the relations between the two kinds of amplitudes.

---

[1]) Since we observe only quadratic expressions of the amplitudes, there are often discrete ambiquities in their reconstruction; the non linear constraints can be used to remove some of these ambiguites.

Table 1

Number of real amplitudes and of polarization observables for usual types of two-body reactions with meson beam and with unpolarized or polarized target. The reactions whose numbers are boxed will be studied in the text

| Type of Reaction | Amplitudes | Observables: Unpolarized Target $U$ | Observables: Polarized target, Transversal $T$ | Observables: Polarized target, Longitudinal $L$ |
|---|---|---|---|---|
| $\pi p \to \pi N \left(0, \frac{1^e}{2}\right)$ | 3 | 1 | 2 | +0 |
| $K\Lambda \left(0, \frac{1}{2}\right)$ | 3 | 2 | $3 + 3$ | +2 |
| $\rho N \left(1^e, \frac{1^e}{2}\right)$ | 11 | 4 | 10 | +2 |
| $K^*\Lambda \left(1^e, \frac{1}{2}\right)$ | 11 | $10 + 2$ | $11 + 25$ | +12 |
| $A_2N \left(2^e, \frac{1^e}{2}\right)$ | 19 | 9 | $18 + 6$ | +6 |
| $K^{**}\Lambda \left(2^e, \frac{1}{2}\right)$ | 19 | $18 + 12$ | $19 + 71$ | +30 |
| $\left(l^e, \frac{1^e}{2}\right)$ | $4(2l + 1) - 1$ | $(l + 1)^2$† | $(l + 1)(3l + 2)$† | $+l(l + 1)$ |
| $\left(l^e, \frac{1}{2}\right)$ | $4(2l + 1) - 1$ | $2(l + 1)(2l + 1)$† | $6(l + 1)(2l + 1)$ | $+2(l + 1)(2l + 1)$ |
| $\pi p \to \pi\Delta \left(0, \frac{3^e}{2}\right)$ | 7 | 4 | $7 + 3$ | +2 |
| $K\Sigma^* \left(0, \frac{3}{2}\right)$ | 7 | $6 + 2$ | $7 + 17$ | +8 |
| $\rho\Delta \left(1^e, \frac{3^e}{2}\right)$ | 23 | 20 | $23 + 33$ | +16 |
| $K^*\Sigma^* \left(1^e, \frac{3}{2}\right)$ | 23 | $22 + 26$ | $23 + 121$ | +48 |
| $\left(l^e, \frac{3^e}{2}\right)$ | $8(2l + 1) - 1$ | $2(l + 1)(3l + 2)$† | $2(l + 1)(9l + 5)$ | $+2(l + 1)(3l + 1)$ |
| $\left(l^e, \frac{3}{2}\right)$ | $8(2l + 1) - 1$ | $8(l + 1)(2l) + 1)$† | $24(l + 1)(2l + 1)$ | $+8(l + 1)(2l + 1)$ |
| $(l^e, j^e)$ | $2(2l + 1) \times (2j + 1) - 1$ | $(l + 1)\left(j + \frac{1}{2}\right) \times \left(2lj + j + \frac{1}{2}\right)$† | $(l + 1)\left(j + \frac{1}{2}\right) \times \left(6lj + 3j + \frac{1}{2}\right)$†† | $(l + 1)\left(j + \frac{1}{2}\right) \times \left(2lj + j - \frac{1}{2}\right)$ |
| $(l^e, j)$ | $2(2l + 1) \times (2j + l) - 1$ | $(l + 1)\left(j + \frac{1}{2}\right) \times (2l + 1)(2j + 1)$ | $(l + 1)\left(j + \frac{1}{2}\right) \times 3(2l + 1)(2j + 1)$ | $(l + 1)\left(j + \frac{1}{2}\right) \times (2l + 1)(2j + 1)$ |

† One amplitude is ghost.
†† One amplitude is ghost for $j = 1/2$.

Table I lists only general reaction types. In order to appreciate how many usual reactions correspond to each type, we present in Tables 2 and 3 several lists of such reactions. We have considered only reactions with practicable beam and targets and with well classified final particles (nonets or decuplets). We obtain in this way a list of 252 reactions belonging to the 8 types set in boxes in table 1 and whose amplitude reconstruction will be treated in detail below.

Furthermore in table 2 and 3 the isospin relations between these reactions are explicitly given. For reactions related via one isospin channel, the ratio of their amplitudes to the isospin amplitudes are fixed coefficients which are given as the coefficients of $(f)$ in the brackets. Then the measurement of the amplitudes for only one reaction yields the amplitudes of all reactions in the same column. For reaction related via two isospin channels the amplitudes satisfy the relations indicated in part c) of each table. In these cases the reconstruction of amplitudes for two independent reactions, because of the overall phase and eventually of the ghost phase, does not fix the amplitudes of a third reaction. But this reconstruction for three (two by two linearly independent) reactions fixes the amplitudes of all the reactions in the same column[2]) up to an overall and eventually one ghost phases. The ghost phase of the whole set of reactions can be fixed by an experiment with polarized target for only one reaction of the set.

Sections 2 and 3 expose the general tools for the measurement of observables. In section 4 the concrete recipes for the amplitude reconstruction in each type of reaction are given. The hurried reader can directly skip to the reaction type in section 4 he is interested in.

Table 2

Listing of 140 reactions of types $\pi p \to K\Sigma^*$, $K^*\Lambda$, $K^{**}\Lambda$, $K^*\Sigma^*$ with their isospin relations. Their transversity amplitudes can be reconstructed with unpolarized target up to one phase (which can be measured with either longitudinally or transversally polarized target)

a) 30 reactions of type $K\Sigma^*$†. Similar reactions of type $K^*\Sigma^*$ are obtained by changing $(\pi\eta\eta' K) \to (\rho\omega\varphi K^*)$††.

| $\pi N \to$ | K | $\Sigma^*$ | |
|---|---|---|---|
| $\pi^+ p$ | + | + | $(a_1)$ |
| $\pi^- p$ | +<br>0 | −<br>0 | $(a_2)$<br>$(a_3)$ |
| $\pi^+ n$ | +<br>0 | 0<br>+ | $(a_3)$<br>$(a_2)$ |
| $\pi^- n$ | 0 | − | $(a_1)$ |

| $\bar{K}N \to$ | $\eta(\eta')\ \Sigma^*$ | $\pi\Sigma^*$ | $K\ \Xi^*$ |
|---|---|---|---|
| $K^- p$ | 0 0 $(f)$ | + − $(b_4)$<br>0 0 $(b_3)$<br>− + $(b_2)$ | + − $(c_1)$<br>0 0 $(c_2)$ |
| $K^- n$ | 0 $-(\sqrt{2}\, f)$ | 0 − $(-b_1)$<br>− 0 $(b_1)$ | 0 − $(c_3)$ |
| $\bar{K}^0 p$ | 0 $+(\sqrt{2}\, f)$ | + 0 $(-b_1)$<br>0 + $(b_1)$ | + 0 $(c_3)$ |
| $\bar{K}^0 n$ | 0 0 $(f)$ | + − $(-b_2)$<br>0 0 $(-b_3)$<br>− + $(-b_4)$ | + − $(-c_1)$<br>0 0 $(c_2)$ |

† For their measurement and amplitude reconstruction, cf. tables 5—6.
†† For their measurement and amplitude reconstruction, cf. table 13.

[2]) The amplitude reconstruction for all other reactions in the column supplies checks of isospin invariance, and even for only three reactions, when several ampltudes with their relative phases are reconstructed.

If its experiment uses polarized target we advise him to read section 4.1 in order to see the connection of the terminology with the standard Wolfenstein parameters. For all other necessary tools he will find references to section 2 and 3.

Table 2 — cont'd

b) 40 reactions of type $K^*\Lambda$††† . Similar reactions of type $K^{**}\Lambda$ are obtained by changing $(\rho\omega\varphi K^*) \to (A_2 f f' K^{**})$††††.

| $\pi N \to$ | $K^*\Lambda$ | | | $K^*\Sigma$ | | |
|---|---|---|---|---|---|---|
| $\pi^+ p$ | | | | + | + | $(a_1)$ |
| $\pi^- p$ | 0 | 0 | $(f)$ | + | − | $(a_2)$ |
| | | | | 0 | 0 | $(a_3)$ |
| $\pi^+ n$ | + | 0 | $(-f)$ | + | 0 | $(a_3)$ |
| | | | | 0 | + | $(a_2)$ |
| $\pi^- n$ | | | | 0 | − | $(a_1)$ |

| $\bar{K}N \to$ | $\omega(\varphi)\,\Lambda$ | | | $\rho\ \Lambda$ | | | $\omega(\varphi)\,\Sigma$ | | | $\rho\ \Sigma$ | | | $K^*\ \Xi$ | | |
|---|---|---|---|---|---|---|---|---|---|---|---|---|---|---|---|
| $K^- p$ | 0 | 0 | $(f)$ | 0 | 0 | $(f)$ | 0 | 0 | $(f)$ | + | − | $(b_4)$ | + | − | $(c_1)$ |
| | | | | | | | | | | 0 | 0 | $(b_3)$ | 0 | 0 | $(c_2)$ |
| | | | | | | | | | | − | + | $(b_2)$ | | | |
| $K^- n$ | | | | − | 0 | $(\sqrt{2}\,f)$ | 0 | − | $(\sqrt{2}\,f)$ | 0 | − | $(-b_1)$ | 0 | − | $(c_3)$ |
| | | | | | | | | | | − | 0 | $(b_1)$ | | | |
| $\bar{K}^0 p$ | | | | + | 0 | $(\sqrt{2}\,f)$ | 0 | + | $(\sqrt{2}\,f)$ | + | 0 | $(-b_1)$ | + | 0 | $(c_2)$ |
| | | | | | | | | | | 0 | + | $(b_1)$ | | | |
| | | | | | | | | | | + | − | $(-b_2)$ | | | |
| $\bar{K}^0 n$ | 0 | 0 | $(-f)$ | 0 | 0 | $(f)$ | 0 | 0 | $(f)$ | 0 | 0 | $(-b_3)$ | + | − | $(c_2)$ |
| | | | | | | | | | | − | + | $(-b_4)$ | 0 | 0 | $(c_1)$ |

c) Triangular isospin relations between the amplitudes of the reactions in a) and b)

$$a_1 = a_2 + \sqrt{2}\,a_3 \qquad \sqrt{2}\,b_1 = 2(b_2 + b_3) = b_2 - b_4 = -2(b_3 + b_4) \qquad c_1 + c_2 = c_3$$

††† For their measurement and amplitude reconstruction, cf. tables 8—9.
†††† For their measurement and amplitude reconstruction, c.f. table 11.

Table 3

Listing of 112 reactions of types $\pi p \to \pi\Delta$, $\rho N$, $A_2N$, $\rho\Delta$ with their isospin relations. Their transversity amplitudes can be reconstructed completely or up to one phase) with transversally polarized target

a) 34 reactions of type $\pi\Delta$†. Similar reactions of type $\rho\Delta$ are obtained by changig $(\pi\eta\eta'K) \to \rho\omega\varphi K^*$)††
In both cases the amplitudes can be completely reconstructed.

| $\pi N\uparrow\to$ | $\pi$ | $\Delta$ | | $\eta(\eta')$ | $\Delta$ | |
|---|---|---|---|---|---|---|
| $\pi^+p$ | + | + | $(-\sqrt{2}\,d_1)$ | 0 | ++ | $(\sqrt{3}\,f)$ |
| | 0 | ++ | $(\sqrt{3}\,d_1)$ | | | |
| | + | − | $(d_2)$ | | | |
| $\pi^-p$ | 0 | 0 | $(d_3)$ | 0 | 0 | $(f)$ |
| | − | + | $(d_4)$ | | | |
| | + | 0 | $(-d_4)$ | | | |
| $\pi^+n$ | 0 | + | $(-d_3)$ | 0 | + | $(f)$ |
| | − | ++ | $(-d_2)$ | | | |
| $\pi^-n$ | 0 | − | $(-\sqrt{3}\,d_1)$ | | | |
| | − | 0 | $(\sqrt{2}\,d_1)$ | 0 | − | $(\sqrt{3}f)$ |

| $KN\uparrow\to$ | K | $\Delta$ | |
|---|---|---|---|
| $K^+p$ | + | + | (f) |
| | 0 | ++ | $(-\sqrt{3}f)$ |
| $K^+n$ | + | 0 | $(f)$ |
| | 0 | + | $(-f)$ |
| $K^0p$ | + | 0 | $(f)$ |
| | 0 | + | $(-f)$ |
| $K^0n$ | + | − | $(\sqrt{3}f)$ |
| | 0 | 0 | $(-f)$ |

| $\overline{K}N\uparrow\to$ | $\overline{K}$ | $\Delta$ | |
|---|---|---|---|
| $K^-p$ | 0 | 0 | $(f)$ |
| | − | + | $(-f)$ |
| $K^-n$ | 0 | − | $(\sqrt{3}\,f)$ |
| | − | 0 | $(-f)$ |
| $\overline{K}^0p$ | 0 | + | $(f)$ |
| | − | ++ | $(-\sqrt{3}\,f)$ |
| $\overline{K}^0n$ | 0 | 0 | $(f)$ |
| | − | + | $(-f)$ |

† For their measurement and amplitude reconstuuction, cf. table 7.
†† For their measurement and amplitude reconstruction, c.f. table 14.

Table 3 — cont'd

b) 22 reactions of type $\rho N$††† . Similar reactions of type $A_2N$ are obtained by changing $(\rho\omega\varphi K^*) \to (A_2\, f f' K^{**})$††††. In both cases the transversity amplitudes can be reconstructed up to one phase.

| $\pi N \uparrow \to$ | $\rho$ | $N$ | | $\omega(\varphi)$ | $N$ | |
|---|---|---|---|---|---|---|
| $\pi^+p$ | + | + | $(a_1)$ | | | |
| $\pi^-p$ | 0<br>− | 0<br>+ | $(a_3)$<br>$(a_2)$ | 0 | 0 | $(f)$ |
| $\pi^+n$ | +<br>0 | 0<br>+ | $(a_2)$<br>$(a_3)$ | 0 | + | $(-f)$ |
| $\pi^-n$ | − | 0 | $(a_1)$ | | | |

| KN ↑ → | K* | N | |
|---|---|---|---|
| K⁺p | + | + | $(c_3)$ |
| Kn | +<br>0 | 0<br>+ | $(c_2)$<br>$(c_1)$ |
| K⁰p | +<br>0 | 0<br>+ | $(c_1)$<br>$(c_2)$ |
| $\bar{K}^0$n | 0 | 0 | $(c_3)$ |

| $\bar{K}N \uparrow \to$ | $\bar{K}^*$ | N | |
|---|---|---|---|
| K⁻p | 0<br>− | 0<br>+ | $(c_1)$<br>$(c_2)$ |
| K⁻n | − | 0 | $(c_3)$ |
| $\bar{K}^0$p | 0 | + | $(c_3)$ |
| $\bar{K}^0$n | 0<br>− | 0<br>+ | $(c_2)$<br>$(c_1)$ |

c) Triangular isospin relations between the amplitudes of the reactions in a) and b)

$$a_1 = a_2 + \sqrt{2a_3} \qquad c_1 + c_2 = c_3 \qquad d_1 = -\sqrt{\frac{2}{3}}\, d_2 - d_3 = -\sqrt{\frac{1}{6}}\, d_2 + \sqrt{\frac{1}{2}}\, d_4 = d_3 + \sqrt{2}\, d_4$$

††† For their measurement and amplitude reconstruction, cf. table 10.
†††† For their measurement and amplitude reconstruction, cf. table 12.

## 2. Some Basic Tools of Quasi Two Body Reaction Analysis

In this section we present some well known features of the formalism used in the study of quasi two body reactions. For more details one refers to [*10, 11, 12, 13*].
Let us first precise our notations. A quasi two body reaction will be denoted by

$$1 + 2 \to 3 + 4.$$

1 denotes the beam, 2 the target, and we call 3 and 4 respectively the particles which share some physical properties with the beam and the target (for instance 1 and 3 are mesons, 2 and 4 are baryons) so that the $t$- and $u$-channels are well defined. The 4-momenta of the particles are denoted by $p_i (i = 1, 2, 3, 4)$ with $p_1 + p_2 = p_3 + p_4$, their spins by $j_i$ and their masses by $m_i$ $(m_i \neq 0)$.

## 2.1. Covariant quantization systems

To describe the polarization of the initial and final states one must fix a quantization frame for each spinning particle. Several different choices are possible, the most popular are the helicity and transversity frames in the $s$-, $t$- and $u$-channels[3]). Unfortunately there is not as yet a universal agreement on the definition of these quantization frames; the most usual conventions are the following.

i) For each particle and for each channel, the transversity quantization axis ${}^{T}n^{(3)}$ and the helicity second axis ${}^{H}n^{(2)}$ are along the "Basel normal" $n$ to the reaction plane, defined by

$$n \cdot p_i = 0 \quad (i = 1, 2, 3), \qquad n^2 = -1, \quad \det(n, p_1, p_2, p_3) > 0 \tag{2.1}$$

where the last condition is equivalent to $\boldsymbol{n} \cdot \boldsymbol{p}_1 \times \boldsymbol{p}_3 > 0$ in the laboratory system, or in the center of mass system.

ii) For each particle and for each channel the helicity and the transversity frames have the same first axis, ${}^{T}n^{(1)} = {}^{H}n^{(1)}$.
Note that with these two conventions, the transversity second axis ${}^{T}n^{(2)}$ and the helicity quantization axis ${}^{H}n^{(3)}$ have oposite directions.[4]) Furthermore the transversity frame is transformed into the helicity frame by a rotation[5]) $\tilde{R} = (-\pi/2, \pi/2, \pi/2)$ of $+\pi/2$ around the common axis $n^{(1)}$. The unitary representations $D^j(\tilde{R})$ of this rotation have several useful properties (cf. Ref. [*13*], [*14*]).

iii) For each particle $i$ ($i = 1, 2, 3, 4$) and for each channel $a$ ($a =$ s, $t$, $u$), the helicity quantization axis ${}^{H}_{a}n_i{}^{(3)}$ and the transversity second axis ${}^{T}_{a}n_i{}^{(2)}$ are defined by

$${}^{H}_{a}n_i{}^{(3)} = -{}^{T}_{a}n_i{}^{(2)} = \varepsilon q_i(a), \qquad \varepsilon = \pm 1, \tag{2.2}$$

with

$$q_i(a) = [\sinh \phi_i(a)]^{-1} (\hat{p}_i \cosh \phi_i(a) - \hat{p}_{ai}) \tag{2.3}$$

where $\hat{p}_i = p_i/m_i$, $\cosh \phi_i(a) = \hat{p}_i \cdot \hat{p}_{ai}$, $\sinh \phi_i(a) > 0$, and where $ai$ is the particle associated to particle $i$ in the channel $a$, i.e., for $i = (1, 2, 3, 4)$, $si = (2, 1, 4, 3)$, $ti = (3, 4, 1, 2)$, $ui = (4, 3, 2, 1)$[6]).

iv) The sign $\varepsilon$ in eq. (2.2) is a last convention to be chosen. JACOB and WICK [*17*] and COHEN-TANNOUDJI, MOREL and NAVELET [*18*] define $\varepsilon = +1$ so that the $s$-helicity quantization axis of particle $i$, ${}^{H}_{s}\boldsymbol{n}_i{}^{(3)}$, be along the 3-momentum $\boldsymbol{p}_i$ of this particle, in the center of mass system ($\boldsymbol{p}_1 + \boldsymbol{p}_2 = 0 = \boldsymbol{p}_2 + \boldsymbol{p}_4$). On the contrary, Gottfried and JACKSON [*19*] define $\varepsilon = -1$ so that their $t$-helicity quantization axis ${}^{H}_{t}\boldsymbol{n}_i{}^{(3)}$, in the rest system of particle $i$, is along the 3-monentum $\boldsymbol{p}_{ti}$ of the particle $ti$ associated to $i$ in the $t$-channel.

---

[3]) Cf. refs. [*10*, *11*].

[4]) This means that in any channel the $x$, $y$, $z$ axes are related by $({}^{T}x, {}^{T}y, {}^{T}z) = {}^{H}x, -{}^{H}z, {}^{H}y)$. Ref. [*14*] and the Cracow group use the same convention although a misprint in ref. [*14*] says the contrary. Many other conventions are occasionally used. Ref. [*2*] and [*3*] use $(-{}^{T}x, {}^{T}y, {}^{T}z) = ({}^{H}x, {}^{H}z, {}^{H}y)$. Ref. [*8*] uses $({}^{T}x, {}^{T}y, {}^{T}z) = ({}^{H}z, {}^{H}x, {}^{H}y)$. Ref. [*15*] compares the conventions of ref. [*8*] and that of the text and it recommends the latter.

[5]) We use the Euler angles and the rotation matrices of ROSE [*16*].

[6]) We call these frames the $s$-, $t$-, or $u$-helicity frames and the $s$-, $t$-, or $u$-transversity frames. Some people use the following vocabulary for helicity frames: $t$-helicity frame = Jackson-frame, $s$-helicity frame = helicity frame; and they extend it to transversity frames: $t$-transversity frame = Jackson transversity frame, $s$-transversity frame = helicity transversity frame! We find that this last expression is an awful barbarism since helicity and transversity are two mutually exclusive notions.

For convenience in this paper we quantize the spin of the target in the $s$-channel transversity frame with $\varepsilon = +1$, i.e., in the laboratory system the second axis $\boldsymbol{n}^{(2)}$ is in the direction of the beam 3-momentum $\boldsymbol{p}_1$. Then the longitudinal polarization of the target (with respect to the beam) is in the $y$-direction, and the transverse polarization is in the $(x, z)$ plane with the $z$-direction along the normal to the reaction plane (cf. Fig. 1).
For the final particles we quantize the spins in a transversity frame. We do not precise the channel since all subsequent equations are independent of this choice of channel.

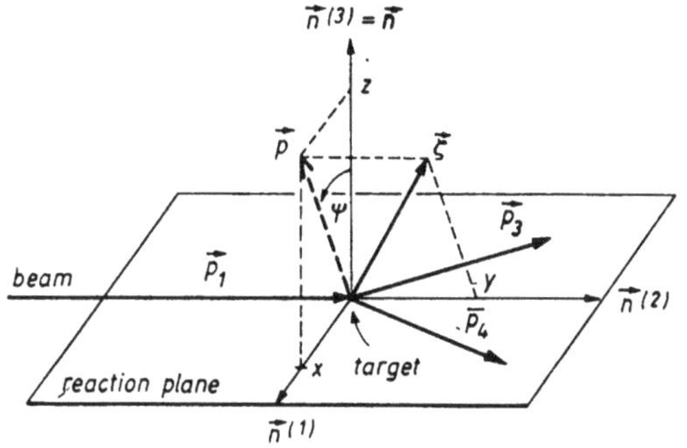


Fig. 1. The $s$-transversity frame for the target quantization (read $\vec{l}$ instead of $\vec{p}$)

## 2.2. Amplitudes

At fixed energy and momentum transfer, the Hilbert space $\mathscr{H}_e = \mathscr{H}_1 \otimes \mathscr{H}_2$ of initial particles has $(2j_1 + 1)(2j_2 + 1)$ dimensions. The transition operator for the reaction is a linear map $T$ between the two spaces. The transversity and helicity amplitudes are the $\mathscr{N} = \prod_{i=1}^{4} (2j_i + 1)$ matrix elements of the transition operator in the transversity and helicity bases. For channel $a$ they are respectively denoted by ${}_aT^{\lambda_3\lambda_4}_{\lambda_1\lambda_2}$ and ${}_aH^{\lambda_3\lambda_4}_{\lambda_1\lambda_3}$. They are related by

$$
{}_aT = [D^{j_3}(\tilde{R}) \otimes D^{j_4}(\tilde{R})]\, {}_aH[D^{j_1}(\tilde{R})^\dagger \otimes D^{j_2}(\tilde{R})^\dagger]. \tag{2.4}
$$

The reflection through the reaction plane, called Bohr-symmetry or B-symmetry, leaves invariant the 4-momenta of the four particles. It acts on the polarization space of a spin-parity $j^\eta$ particle by the operator $B(j) = \eta D^j(n, \pi)$, the product of the parity $\eta$ of the particle times the unitary representation of the rotation by $\pi$ around the normal to the reaction plane. If parity is conserved in the reaction, the transition operator is invariant by B-symmetry i.e.

$$
[B(j_3) \otimes B(j_4)]\, T[B(j_1)^\dagger \otimes B(j_2)^\dagger] = T, \tag{2.5}
$$

and the matrix elements of $T$ satisfy the relations

$$
\eta(-1)^{\lambda_3+\lambda_4-\lambda_1-\lambda_2}\, T^{\lambda_3\lambda_4}_{\lambda_1\lambda_2} = T^{\lambda_3\lambda_4}_{\lambda_1\lambda_2} \tag{2.6a}
$$

$$
\eta(-1)^{j_3-\lambda_3+j_4-\lambda_4+j_1-\lambda_1+j_2-\lambda_2}\, H^{-\lambda_3-\lambda_4}_{-\lambda_1-\lambda_2} = H^{\lambda_3\lambda_4}_{\lambda_1\lambda_2} \tag{2.6b}
$$

where $\eta$ is the relative parity of the particles, i.e., $\eta = \eta_1\eta_2\eta_3\eta_4$ with $\eta_i =$ parity of particle $i$.

In the collinear case of forward or backward scattering, the transition operator must be invariant under rotation around the reaction axis. This imposes "collinearity constraints" among its matrix elements, which are most simply written in the helicity basis

$$H^{\lambda_3\lambda_4}_{\lambda_1\lambda_2} = 0, \quad \text{for} \quad \lambda_1 + \lambda_2 \neq \lambda_3 + \lambda_4 \tag{2.6c}$$

## 2.3. Density matrices

For a single particle of spin $j$ and fixed 4-momentum, the polarization state is described by a Hermitian, positive, trace one operator acting on the Hilbert space $\mathscr{H}(j)$ and represented by a matrix $\varrho(j)$. For a system of two particles (spins $j$ and $j'$) of fixed 4-mo menta, the polarization operator acts on the space $\mathscr{H}(j) \otimes \mathscr{H}(j')$ and is represented by the joint density matrix $\varrho(j, j')$. When the particles are uncorrelated, this matrix can be written in the form of a tensor product $\varrho(j) \otimes \varrho(j')$.
For later use, especially for the study of decay angular distributions, it is useful to introduce the polarization multipole parameters of these density matrices.

i) The single particle density matrix $\varrho(j)$ is expanded on a set of basis matrices $T(j)^L_M$ $(L = 0, \ldots, 2j;\ M = -L, \ldots, +L)$, the matrix elements of which are Clebsch-Gordan coefficients

$$\left(T(j)^L_M\right)^{\lambda}_{\lambda'} = \langle jL\lambda' M \mid j\lambda \rangle . \tag{2.7}$$

The multipole expansion of $\varrho(j)$ reads

$$\varrho(j) = \frac{1}{2j+1}\left[\mathbf{1} + \sum_{L=1}^{2j} (2L+1) \sum_{M=-L}^{+L} \overline{t^L_M} T(j)^L_M\right], \tag{2.8}$$

the expansion coefficients $t_M{}^L$ are the multipole parameters. We have exhibited the trace of the matrix. We could have written an expansion from $L = 0$ to $2j$, with $T(j)_0{}^0 = 1$ and $t_0{}^0 = 1$.

ii) A similar multipole expansion can be written for the joint density matrix $\varrho(j, j')$. The set of basis matrices is the tensor products $T(j)^L_M \otimes T(j')^{L'}_{M'}$ and the corresponding multipole parameters are denoted by $t^{LL'}_{MM'}$. The expansion is

$$\varrho(j, j') = \sum_{L=0}^{2j} \sum_{L'=0}^{2j'} \frac{(2L+1)(2L'+1)}{(2j+1)(2j'+1)} \sum_{M=-L}^{+L} \sum_{M'=-L'}^{+L'} \overline{t^{LL'}_{MM'}} T(j)^L_M \otimes T(j)^{L'}_{M'} . \tag{2.9}$$

Note that the multipole parameters $t(j)_M{}^L$ and $t(j')^{L'}_{M'}$ of the single particle density matrices[7] $\varrho(j) = \mathrm{tr}_{j'}\, \varrho(j, j')$ and $\varrho(j') = \mathrm{tr}_j\, \varrho(j, j')$ are $t(j)^L_M = t^{L0}_{M0}$ and $t(j')^{L'}_{M'} = t^{0L'}_{0M'}$.
A density matrix $\varrho$ can be split into a B-symmetric part $\varrho^B$ and a B-antisymmetric part $\varrho^A$ satisfying the conditions

$$B\varrho^B B^\dagger = \varrho^B, \qquad B\varrho^A B^\dagger = -\varrho^A .$$

i) For a single particle, in transversity quantization, the matrix elements of $\varrho^B$ and $\varrho^A$ satisfy the relations

$$(\varrho^B)^{\lambda}_{\lambda'} = (-1)^{\lambda-\lambda'} (\varrho^B)^{\lambda}_{\lambda'}, \qquad (\varrho^A)^{\lambda}_{\lambda'} = -(-1)^{\lambda-\lambda'} (\varrho^A)^{\lambda}_{\lambda'} \tag{2.10}$$

---

[7]) $\mathrm{tr}_j$ (and $\mathrm{tr}_{j'}$) represent the partial trace in the space $\mathscr{H}(j)$ (and $\mathscr{H}(j')$), e.g., $\varrho(j)_\nu{}^\mu = (\mathrm{tr}_{j'}\, \varrho(j, j'))_\nu{}^\mu = \sum_\lambda \varrho(j, j')_{\nu\lambda}{}^{\mu\lambda}$.

and the multipole parameters satisfy the conditions

$$t^{LB}_M = (1)^M \, t^{LB}_M, \qquad t^{LA}_M = -(-1)^M \, t^{LA}_M. \tag{2.11}$$

ii) For a joint density matrix, in transversity quantization, the matrix elements satisfy

$$(\varrho^{B,A})^{\lambda\mu}_{\lambda'\mu'} = \pm(-1)^{\lambda+\mu-\lambda'-\mu'} \, (\varrho^{B,A})^{\lambda\mu}_{\lambda'\mu'}, \begin{cases} + \text{ for } B \\ - \text{ for } A \end{cases} \tag{2.12}$$

and the joint multipole parameters satisfy

$$(t^{LL'}_{MM'})^{B,A} = \pm(-1)^{M+M'} \, (t^{LL'}_{MM'})^{B,A} \begin{cases} + \text{ for } B \\ - \text{ for } A. \end{cases} \tag{2.13}$$

In a quasi two body reaction with spin zero beam and spin 1/2 target, the transition matrix has 2 columns ($T^{\lambda_3\lambda_4}_{0\pm 1/2}$). If the target is unpolarized, the density matrix of the final particles, at fixed energy and momentum transfer, is obtained by[8])

$$\sigma\varrho_f = \frac{1}{2} \, TT^\dagger \tag{2.14}$$

(where $\sigma$ is the differential cross section) and $\varrho_f$ has rank 2. If furthermore $T$ is B-symmetric, $\varrho_f$ is B-symmetric too and in transversity quantization it can be written in the form of a direct sum $\varrho_f = \varrho_1 \oplus \varrho_2$, each of which has rank one. This property enforces very strong relations between the matrix elements or the multipole parameters of $\varrho_f$.

### 2.4. Decay angular distributions

If the final particles of the quasi two body reaction are unstable, the angular distribution of their decay products (and occasionally the cascade angular distribution) provides some information on their polarization state. In this paper we limit ourselves to two body decays, however we give some results for 3 body decays in sect. 2.4.4.

#### 2.4.1. Single decay angular distribution

Assume first that only one of the final particles is unstable and let $j$ be the spin of this particle ($j = j_3$ or $j = j_4$). In the rest system of this particle the kinematics of the decay is determined by the polar angle $\theta$ and azimuthal angle $\phi$ of one of the decay products with respect to the quantization frame of the decaying particle. If $M$ is the decay operator, the normalized angular distribution is defined by

$$I(\theta, \phi) = \text{tr}\, M\varrho(j)\, M^\dagger \,/ \int \text{tr}\, M\varrho(j)\, M^\dagger d(\cos\theta)\, d\phi. \tag{2.15}$$

This angular distribution is linear in the multipole parameters, it can be written

$$I(\theta, \phi) = \frac{1}{4\pi} + \sum_{L=1}^{2j} C(L) \sum_{M=-L}^{+L} \overline{t^L_M} \, Y^L_M(\theta, \phi) \tag{2.16}$$

where the coefficients $C(L)$ depend on the spins of the decay products and on the dynamics of the decay. If parity is conserved in the decay (e.g. $\rho \to \pi\pi \quad \Delta \to N\pi$) the coeffi-

[8]) This relation fixes the normalization of $T$.

cients $C(L)$ vanish for $L =$ odd; the angular distribution is an analyser of the even polarization. If parity is violated in the decay (e.g. $\Lambda \to p\pi$, $\Omega \to \Xi\pi$) no $C(L)$ coefficient vanishes; the angular distribution is an analyser of the complete polarization. In many cases angular momentum conservation implies that only one amplitude contributes to the decay. Then the coefficients $C(L)$ are pure numbers. If two amplitudes contribute they depend on one dynamical parameter. Here are the values of the coefficients $C(L)$ for some usual decays

$$\begin{aligned}
&1^- \to 0^-0^- : \sqrt{4\pi}\, C(2) = -\sqrt{2},\\
&2^+ \to 0^-0^- : \sqrt{4\pi}\, C(2) = -\sqrt{10/7}, \qquad \sqrt{4\pi}\, C(4) = \sqrt{18/7},\\
&2^+ \to 1^-0^- : \sqrt{4\pi}\, C(2) = -\sqrt{5/14}, \qquad \sqrt{4\pi}\, C(4) = -\sqrt{8/7},\\
&\frac{1}{2} \to \frac{1}{2}\, 0 : \sqrt{4\pi}\, C(1) = \alpha,\\
&\frac{3^+}{2} \to \frac{1^+}{2}\, 0^- : \sqrt{4\pi}\, C(2) = -1,
\end{aligned} \tag{2.17}$$

where $\alpha$ is the asymmetry parameter of the parity violating decay $1/2 \to 1/2\ 0$. With these known values of the $C(L)$ coefficients, the $t^L_M$ are obtained by a maximum likelihood analysis of the angular distribution, or by a moment analysis which yields $\big(\Omega = (\theta, \phi),\ d\Omega = d(\cos\theta)\, d\phi\big)$

$$C(L)\, t^L_M = \langle Y^L_M(\Omega)\rangle \equiv \int I(\Omega)\, Y^L_M\, d\Omega. \tag{2.18}$$

Experimentally the moments $\langle Y^L_M(\Omega)\rangle$ of the angular distribution are the values of the spherical harmonics $Y^L_M(\theta, \phi)$ for all events in an ensemble of fixed energy and momentum transfer

$$\langle Y^L_M(\Omega)\rangle = \frac{1}{N} \sum_{i=1}^{N} Y^L_M(\theta_i, \phi_i)$$

where the index $i$ specifies the event which is considered, and $N$ is the total number of events in the ensemble.

### 2.4.2. Joint decay angular distribution

If both final particles of the quasi two body reaction are unstable, one may study the correlations between the directions of their decay products. Let $j$ and $j'$ be the spins of the final particles. The joint decay angular distribution reads

$$I(\theta, \phi; \theta', \phi') = \sum_{L=0}^{2j} \sum_{L'=0}^{2j'} C(L)\, C'(L') \sum_{M=-L}^{+L} \sum_{M'=-L'}^{+L'} \overline{t^{LL'}_{MM'}}\, Y^L_M(\theta, \phi)\, Y^{L'}_{M'}(\theta', \phi') \tag{2.19}$$

where $C(L)$ and $C'(L')$ are the coefficients of the single decays, with $C(0) = C'(0) = 1/\sqrt{4\pi}$. These coefficients being known, the parameters $t^{LL'}_{MM'}$ are obtained by a best fit analysis of the angular distribution or by a moment analysis $\big(\Omega = (\theta, \phi),\ \Omega' = (\theta, \phi'),\ d\Omega = d(\cos\theta)\, d\phi,\ d\Omega' = d(\cos\theta')\, d\phi'\big)$

$$C(L)\, C'(L')\, t^{LL'}_{MM'} = \langle Y^L_M(\Omega)\, Y^{L'}_{M'}(\Omega')\rangle \equiv \int I(\Omega, \Omega')\, Y^L_M(\Omega)\, Y^{L'}_{M'}(\Omega')\, d\Omega\, d\Omega'. \tag{2.20}$$

### 2.4.3. Cascade decay angular distribution

Consider the cascade decay $C \to A + B$, $A \to A_1 + B_1$ (with spins $j(C) = j$, $j(A) = j(A_1) = 1/2$, $j(B) = j(B_1) = 0$); the first decay is parity conserving and the second decay is parity violating. (e.g. $\Sigma^* \to \Lambda\pi$, $\Lambda \to p\pi$). We denote by $\theta$ and $\phi$ the angles of $A$ with respect to the quantization frame of $C$, and by $\theta_1$ and $\phi_1$ the angles of $A_1$ with respect to the canonical quantization frame for $A$, deduced from the quantization frame of $C$ by a pure Lorentz transformation (boost). Then the cascade angular distribution is

$$I(\theta, \phi; \theta_1, \phi_1) = \sum_{L=0}^{2j} \sum_{L_1=0}^{1} \sum_{J\,\text{even}} C(L, L_1, J) \sum_{M,N,M_1} \langle J L_1 N M_1 \mid L M \rangle \times \overline{t^L_M}\, Y^J_N(\theta, \phi)\, Y^{L_1}_{M_1}(\theta_1, \phi_1). \quad (2.21\text{a})$$

Instead of the canonical quantization frame for particle $A$ one may use the helicity frame, deduced from the previous one by the rotation $R(\phi, \theta, 0)$. We denote by $\theta_1{}^h$, $\phi_1{}^h$ the angles of $A_1$ with respect to this frame, then the cascade angular distribution is

$$I(\theta, \phi, \theta_1{}^h, \phi_1{}^h) = \sum_{L=0}^{2j} \sum_{L_1=0}^{1} C^h(L_1\ L_1, M_1) \sum_{M,M_1} \overline{t^L_M} \sqrt{\frac{2L+1}{4\pi}}\, D^L(\phi, \theta, 0)^{M_1}_{M}\, Y^{L_1}_{M_1}(\theta_1{}^h, \phi_1{}^h). \quad (2.21\text{b})$$

The coefficients $C(L, L_1, J)$ in Eq. (2.21a) and $C^h(L, L_1, M_1)$ in Eq. (2.21b) depend on the spins and parities of the particle and on the dynamics of the decays, if these involve more than one amplitude.
The most usual decay of this type is $3^+/2 \to 1^+/2\, 0^-$, $1^+/2 \to 1^+/2\, 0^-$. For this cascade decay the non vanishing coefficients $C(L, L_1, J)$ and $C^h(L, L_1, M_1)$ are

$$4\pi C(0, 0, 0) = 1, \qquad 4\pi C(2, 0, 2) = -1$$
$$4\pi C(1, 1, 0) = \alpha \sqrt{5/9}, \qquad 4\pi C(1, 1, 2) = \alpha \sqrt{2/45} \quad (2.22\text{a})$$
$$4\pi C(3, 1, 2) = -\alpha \sqrt{7/5},$$

$$(4\pi)\, C^h(0, 0, 0) = 1, \qquad (4\pi)\, C^h(2, 0, 0) = -1$$
$$(4\pi)\, C^h(1, 1, 0) = \alpha \sqrt{1/15}, \qquad (4\pi)\, C^h(1, 1, \pm 1) = \alpha \sqrt{4/15} \quad (2.22\text{b})$$
$$(4\pi)\, C^h(3, 1, 0) = -\alpha \sqrt{3/5}, \qquad (4\pi)\, C^h(3, 1, \pm 1) = -\alpha \sqrt{2/5}$$

where $\alpha$ is the asymmetry parameter of the second decay. With these known values of the coefficients, the $t^L_M$ parameters are deduced by a best fit adjustment of the decay angular distribution or by a moment analysis $\big(\Omega = (\theta, \phi), \Omega_1 = (\theta_1, \phi_1), \Omega_1{}^h = (\theta_1{}^h, \phi_1{}^h)\big)$

$$C(L, L_1, J)\, t^L_M = \sum_{N,M_1} \langle J L_1 N M_1 \mid L M \rangle \langle Y^J_N(\Omega)\, Y^{L_1}_{M_1}(\Omega_1) \rangle, \quad (2.23\text{a})$$

$$C^h(L, L_1, M_1)\, t^L_M = \sqrt{\frac{2L+1}{4\pi}} \sum_{M_1} \langle D^L(\phi, \theta, 0)^M_{M_1}\, Y^{L_1}_{M_1}(\Omega_1{}^h) \rangle. \quad (2.23\text{b})$$

Note that in the above example, since for $L = 1$ one has two non vanishing coefficients, $C(1, 1, 0)$ and $C(1, 1, 2)$ or ($C^h(1, 1, 0)$ and $C^h(1, 1, \pm 1)$) the parameters $t_M{}^1$ can be measured by two different experimental expressions.

### 2.4.4. Three body decays

Some well known unstable resonances undergo 3 body decays, (e.g. $\eta \to \pi^0\pi^0\pi^0$ or $\pi^+\pi^-\pi^0$ $\omega \to \pi^+\pi^-\pi^0$ $\phi \to \pi^+\pi^-\pi^0$). The final state is determined by 5 quantities, often split into the 2 Dalitz plot variables and the 3 angles which fix the orientation of the decay plane. In the rest system of the decaying particle, let us denote by $\theta$ and $\phi$ the angles of the normal to the decay plane with respect to the quantization frame of the decaying particle. Then the angular distribution $I(\theta, \phi)$ can be written in the same form as eq. (2.16) with coefficients $C(L)$ depending on the spins and parities of the particles and on the dynamics of the decay. For the most usual decay, i.e., $1^- \to 0^-0^-0^-$ (e.g. $\omega$ and $\phi$ decays), the non vanishing $C(L)$ coefficient is a pure number

$$1^- \to 0^-0^-0^- : \sqrt{4\pi}\, C(2) = -\sqrt{2}, \tag{2.24}$$

which happens to be equal to the coefficient of the two body decay $1^- \to 0^-0^-$, (cf. eq. (2.17)).

## 3. Quasi Two Body Reactions with Polarized Target

In this section we study the observables of a quasi two body reaction with polarized target. In the particular case of a spin zero beam and a spin 1/2 polarized target we give explicitly the structure of the final state density matrix and of the differential cross section in terms of the initial polarization. We also show how to measure the multipole parameters which describe the *polarization transfer* between the initial and final states.

### 3.1. Observables

Consider a quasi two body reaction $1 + 2 \to 3 + 4$. The beam and the target are prepared independently hence the initial state is described by a (Hermitian, positive, trace one) density matrix which is a tensor product $\varrho_e = \varrho(j_1) \otimes \varrho(j_2)$. On the contrary, the polarizations of the final particles are generally correlated and the final state is described by a joint density matrix $\varrho_f = \varrho(j_3, j_4)$ which cannot generally be written in a tensor product form. These matrices are related through the transition matrix $T$ by

$$\sigma\varrho_f = T\varrho_e T^\dagger. \tag{3.1}$$

We have denoted by $\sigma$ the double differential cross section

$$\sigma \equiv \frac{d\sigma}{dt\, d\psi} = \text{tr}\; T\varrho_e T^\dagger \tag{3.2}$$

where $t$ is the momentum transfer and $\psi$ is, in the laboratory system, the angle between the Basel normal $\boldsymbol{n}$ and a direction $\boldsymbol{l}$, perpendicular to the beam direction, fixed by the initial polarization. If the initial state is unpolarized, $\varrho_e = \mathbf{1}/n_1 \otimes \mathbf{1}/n_2$ with $n_i = (2j + 1)$, the double differential cross section is denoted by $\sigma_0$

$$\sigma_0 \equiv \left.\frac{d\sigma}{dt\, d\psi}\right|_{\varrho_e = \mathbf{1}/n_1 n_2} = \frac{1}{n_1 n_2}\, \text{tr}\; TT^\dagger. \tag{3.3}$$

In this case, the initial state has no preferential direction $\boldsymbol{l}$ in the laboratory and the double differential cross section $\sigma_0$ is isotropic in $\psi$. Then one may consider the simple differential cross section

$$\frac{d\sigma}{dt} = \int_0^{2\pi} \sigma_0 \, d\psi = 2\pi\sigma_0 . \tag{3.4}$$

At fixed energy and momentum transfer, a complete measurement of the reaction includes the measurement of the double differential cross section $\sigma$ and of the joint final density matrix $\varrho_f$, as functions of the initial polarization $\varrho_e$ and of the angle $\psi$. We call *observables* of the reaction the set of quantities which parametrize these functions and can be effectively measured. The measure is obtained by an analysis of the differential cross section and of the combined angular distribution of the normal $\boldsymbol{n}$ and of the decay products of the final particles (cf. section 3.3. below) for different initial polarizations.

## 3.2. Description of the final state when the target is polarized

From now on we assume that the initial state consists simply of beam of spin zero particles and a target of spin 1/2 particles.

### 3.2.1. The initial state

The initial density matrix $\varrho_e$ is a $2 \times 2$ matrix, which in the laboratory system is described by the polarization pseudo vector $\zeta(\zeta^2 \leqq 1)$, also called the Stokes vector. The projection of this vector on a plane perpendicular to the beam fixes the direction $\boldsymbol{l}$ alluded to previously. As we defined it in section 2.1., for $s$-transversity quantization, in the laboratory system the $\boldsymbol{n}^{(2)}$ axis of the target is in the direction of the beam momentum $\boldsymbol{p}_1$, while $\boldsymbol{n}^{(3)}$ is along the Basel normal $\boldsymbol{n}$ to the reaction plane and $\boldsymbol{n}^{(1)}$ is perpendicular to the beam and to this normal. Then the density matrix $\varrho_e$, in $s$-transversity quantization, is

$$\varrho_e = \frac{1}{2} (\mathbf{1} + x\tau_x + y\tau_y + z\tau_z) \tag{3.5}$$

where $\tau_x$, $\tau_y$, $\tau_z$ are the Pauli matrices, and $x$, $y$, $z$ are the projections of the vector $\zeta$ on the $s$-transversity axes $\boldsymbol{n}^{(1)}$, $\boldsymbol{n}^{(2)}$, $\boldsymbol{n}^{(3)}$ respectively. These components can be written

$$\zeta : x = P_T \sin \psi , \qquad y = P_L , \quad z = P_T \cos \psi \tag{3.6}$$

where $\psi$ is the angle between $\boldsymbol{n}$ and $\boldsymbol{l}$ with the sign of $\boldsymbol{n} \times \boldsymbol{l} \cdot \boldsymbol{p}_1$, see fig. 1. By definition $P_T$ is the length of the projection of $\zeta$ on the $(x, z)$ plane; it is the degree of transverse polarization $0 \leqq P_T \leqq 1$. $P_L$ is the projection of $\zeta$ on the beam; it may be positive or negative and its modulus $|P_L|$ is the degree of longitudinal polarization $0 \leqq |P_L| \leqq 1$. Note that $P_T{}^2 + P_L{}^2 = \zeta^2$ is the degree of polarization of the target. It is important to remark that in general the initial state is not B-symmetric. Indeed the matrices $\mathbf{1}$ and $\tau_z$ are B-symmetric in transversity quantization but $\tau_x$ and $\tau_y$ are not. Then, except in the case of normal polarization, $\zeta \simeq \boldsymbol{n}$ the initial state is not invariant by reflection through the reaction plane.

### 3.2.2. The density matrix of the final state

The density matrix $\varrho_f$ computed from eq. (3.1) with the initial state (3.5) is linear in the components of $\zeta$. It can be written in the form

$$\sigma\varrho_f = \sigma_0(\varrho_0 + z\varrho_z + x\varrho_x + y\varrho_y), \tag{3.7}$$

where $\varrho_0$ is the density matrix of the final state when the target is unpolarized ($\zeta = 0$). If parity is conserved in the reaction the transition matrix $T$ is B-symmetric and since the matrices $\mathbf{1}$ and $\tau_z$ are B-symmetric, the matrices $\varrho_0$ and $\varrho_z$ are B-symmetric too and have non vanishing trace while the matrices $\varrho_x$ and $\varrho_y$ are B-antisymmetric and hence traceless. The density matrix $\varrho_0$ has trace 1 whereas the trace of $\varrho_z$ depends on the dynamics of the reaction; this trace, $P_R$, is sometimes called the "reaction polarization"

$$\operatorname{tr}\varrho_0 = 1, \qquad \operatorname{tr}\varrho_z = P_R, \qquad \operatorname{tr}\varrho_x = \operatorname{tr}\varrho_y = 0. \tag{3.8}$$

By definition the 4 matrices $\sigma_0\varrho_\alpha = 1/2T\tau_\alpha T^\dagger(\alpha = 0, x, y, z, \tau_0 = \mathbf{1})$ are not independent of each other. It is easy to show that the so-called "polarization transfer matrix" (cf. Appendix 1 and ref. [20])

$$W = \sum_\alpha \tilde{\tau}_\alpha \otimes \sigma_0\varrho_\alpha = \sigma_0 \begin{array}{|c|c|}\hline \varrho_0 + \varrho_z & \varrho_x + i\varrho_y \\ \hline \varrho_x - i\varrho_y & \varrho_0 - \varrho_z \\ \hline \end{array} \tag{3.9}$$

($\sim$ = transposition in the initial space) must be positive and have rank 1, since $\sigma W$ can be written

$$W = \frac{1}{2}\tilde{T}\tilde{T}^\dagger \tag{3.9'}$$

where $\tilde{T}$ is the column matrix obtained from the transition matrix $T$ by transposition in the initial space, i.e., its elements are

$$\tilde{T}^{\lambda_1\lambda_2\lambda_3\lambda_4} = \tilde{T}^{\lambda_3\lambda_4}_{\lambda_1\lambda_2}.$$

The matrix $W$ is B-symmetric; in transversity it can be written in the form of a direct sum $W = W_1 \oplus W_2$. The line and column indices of $W_1$ satisfy $\lambda_1 + \lambda_2 - \lambda_3 - \lambda_4 =$ even, those of $W_2$ satisfy $\lambda_1 + \lambda_2 - \lambda_3 - \lambda_4 =$ odd. The rank 1 condition on $W$ implies that either $W_2 = 0$ and rank $W_1 = 1$ or $W_1 = 0$ and rank $W_2 = 1$. Which submatrix is nul depends on the relative parity of the particles. From eq. (2.6a), if $\eta = +1$, $W_2 = 0$, if $\eta = -1$, $W_1 = 0$. The nullity of $W_1$ (or $W_2$) yields linear constraints between the elements of $\varrho_0$ and $\varrho_z$ and of $\varrho_x$ and $\varrho_y$, while the rank 1 condition for $W_2$ (or $W_1$) gives quadratic constraints between the elements of all matrices. Often the decay of the final particles does not allow a complete measurement of the matrices $W_1$ and $W_2$. Then one may obtain constraints on the observable parameters by elimination of the unobserved quantities from the previous equations. This elimination keeps the degree of the linear constraints, but it generally raises the degree of the quadratic constraints.

### 3.2.3. Multipole expansions

i) If only the final particle 4 has spin ($j_3 = 0$, $j_4 = j$), the density matrix $\varrho_f$ is a single particle density matrix. Its multipole expansion is (cf. eq. (2.8))

$$\frac{\sigma}{\sigma_0}\varrho_f = \frac{1}{2j+1}\left[(1 + P_R z)\,\mathbf{1} + \sum_{L=1}^{2j}\sum_{M=-L}^{+L}(2L+1)\{\overline{t^L_M} + z\overline{{}^z t^L_M} + x\overline{{}^x t^L_M} + y\overline{{}^y t^L_M}\}\,T(j)^L_M\right], \tag{3.10}$$

where $t_M^L$ and ${}^z t_M^L$ are the multipole parameters of the B-symmetric matrices $\varrho_0$ and $\varrho_z$, and ${}^x t_M^L$ and ${}^y t_M^L$ are the parameters of the B-antisymmetric matrices $\varrho_x$ and $\varrho_y$. We have exhibited the trace $(1 + P_R z)$ of the matrix $\sigma/\sigma_0\, \varrho_f$; we could keep this term inside the summation $(L = 0, 2j)$, with the conventions $t_0{}^0 = 1$, ${}^z t_0{}^0 = P_R$.

ii) If both final particles 3 and 4 have spin $(j_3 = j,\ j_4 = j')$, the density matrix $\varrho_f$ is a joint density matrix $\varrho_f(j, j')$. Its multipole expansion is (cf. eq. (2.9)).

$$\frac{\sigma}{\sigma_0}\, \varrho_f = \sum_{L=0}^{2j} \sum_{L'=0}^{2j'} \frac{(2L+1)(2L'+1)}{(2j+1)(2j'+1)} \sum_{M=-L}^{+L} \sum_{M'=-L'}^{+L'} \{\overline{t_{MM'}^{LL'}} + z\overline{{}^z t_{MM'}^{LL'}} + x\overline{{}^x t_{MM'}^{LL'}} + Y\overline{{}^y t_{MM'}^{LL'}}\}\, T(j)_M^L \otimes T(j')_{M'}^{L'} \tag{3.11}$$

where $t_{MM'}^{LL'}$, ${}^z t_{MM'}^{LL'}$, ${}^x t_{MM'}^{LL'}$, ${}^y t_{MM'}^{LL'}$ are the multipole parameters of the B-symmetric matrices $\varrho_0$ and $\varrho_z$ and B-antisymmetric matrices $\varrho_x$ and $\varrho_y$ respectively, with the conventions $t_{00}^{00} = 1$ and ${}^z t_{00}^{00} = P_R$.

The multipole parameters ${}^\alpha t_M^L$ and ${}^\alpha t_{MM'}^{LL'}$ $(\alpha = x, y, z)$ are called "polarization transfer" multipole parameters.

## 3.3. Measurement of the observables of a reaction

If the final particles are unstable and undergo two body decays, the final state is characterized by the production angle $\psi$ and by the decay angles $\theta$, $\phi$ and $\theta'$, $\phi'$.

### 3.3.1. The double differential cross sections

From the general form (3.7) of $\sigma\varrho_f$ and from the trace conditions (3.8), one gets

$$\sigma = \operatorname{tr} \sigma\varrho = \sigma_0(1 + P_R z). \tag{3.12}$$

Then from the value (3.6) of the $z$-component of the polarization vector $\zeta$, the $\psi$ dependence of the double differential cross section is

$$\sigma(\psi) = \sigma_0(1 + P_R P_T \cos \psi). \tag{3.13}$$

i) The unpolarized double differential cross section $\sigma_0$ may be obtained by several different ways

$$\sigma_0 = \sigma(\psi)|_{P_T=0} \tag{3.14a}$$

$$\sigma_0 = \sigma\left(\frac{\pi}{2}\right) \tag{3.14b}$$

$$\sigma_0 = \frac{1}{2}\left(\sigma(0) + \sigma(\pi)\right)$$

$$\sigma_0 = \frac{1}{2\pi} \langle \sigma(\psi) \rangle \equiv \frac{1}{2\pi} \int_0^{2\pi} \sigma(\psi)\, d\psi. \tag{3.14d}$$

One may verify that all these ways lead to the same result.

ii) Similarly the asymmetry $P_R P_T$ of the differential cross section can be obtained by several ways. We denote by $\sigma\uparrow$ (resp. $\sigma\downarrow$) the cross sections of the events with the polarization vector $\zeta$ above (resp. under) the reaction plane

$$\sigma\uparrow = \int_{-\pi/2}^{\pi/2} \sigma(\psi)\, d\psi\,, \qquad \sigma\downarrow = \int_{\pi/2}^{3\pi/2} \sigma(\psi)\, d\psi\,, \tag{3.15a}$$

and we use the notation

$$\langle f(\psi)\rangle = \int_0^{2\pi} f(\psi)\, d\psi\,. \tag{3.15b}$$

Then the asymmetry can be obtained by

$$P_R P_T = \frac{\sigma(0) - \sigma(\pi)}{\sigma(0) + \sigma(\pi)} \tag{3.16a}$$

$$P_R P_T = \frac{\pi}{2}\, \frac{\sigma\uparrow - \sigma\downarrow}{\sigma\uparrow + \sigma\downarrow} \tag{3.16b}$$

$$P_R P_T = \frac{\langle \sigma(\psi)\, 2 \cos \psi\rangle}{\langle \sigma(\psi)\rangle}\,. \tag{3.16c}$$

### 3.3.2. Production and single decay angular distribution

Assume first that only one final particle undergo a two body decay (e.g. $\pi p \to \pi\Delta$, $\Delta \to \pi N$; or $\pi p \to \rho N$, $\rho \to \pi\pi$). Then the final state is characterized by 3 angles $\psi$, $\theta$, $\phi$. Les us call $T$ the reaction transition matrix and $M$ the decay transition matrix. The normalized combined angular distribution is defined by

$$I(\psi; \theta, \phi) = \frac{\operatorname{tr} M T \varrho_e T^\dagger M^\dagger}{\int \operatorname{tr} M T \varrho_e T^\dagger M^\dagger \, d(\cos\theta)\, d\phi\, d\psi}\,. \tag{3.17}$$

From eq. (3.1) this can be written

$$I(\psi; \theta, \phi) = \frac{\operatorname{tr} M \dfrac{\sigma}{\sigma_0} \varrho_f M^\dagger}{\displaystyle\int \operatorname{tr} M \frac{\sigma}{\sigma_0} \varrho_f M^\dagger \, d(\cos\theta)\, d\phi\, d\psi}\,. \tag{3.18}$$

Then, by comparison of the multipole expansion of $\sigma/\sigma_0\, \varrho_f$ (cf. eq. (3.10)) with the expansion (2.8) and from the usual decay angular distribution (2.16) one gets the combined normalized angular distribution

$$I(\psi; \theta, \phi) = \frac{1}{2\pi} \left[ \frac{1 + P_R P_T \cos\psi}{4\pi} + \sum_{L=1}^{2j} C(L) \sum_{M=-L}^{+L} \left\{ \overline{t^L_M} + P_T \cos \psi\, \overline{{}^z t^L_M} \right.\right.$$
$$\left.\left. + P_T \sin \psi\, \overline{{}^x t^L_M} + P_L \overline{{}^y t^L_M} \right\} Y^L_M(\theta, \phi) \right] \tag{3.19}$$

where the coefficients $C(L)$ are defined in section 2.4 and are, for the most usual decays, well known numerical coefficients (cf. eq. (2.17)). By inspection of this expression one sees that it allows the measure of the quantities $P_R P_T$, $t^L_M + P_L {}^y t^L_M$, $P_T {}^z t^L_M$, $P_T {}^x t^L_M$, either by a best fit adjustment or by a moment analysis which yields

$$P_R P_T = \langle 2\cos\psi\rangle \equiv \int I(\psi;\Omega)\, 2\cos\psi\, d\Omega\, d\psi \tag{3.20a}$$

$$C(L)\,(t^L_M + P_L{}^y t^L_M) = \langle Y^L_M(\Omega)\rangle \equiv \int I(\psi;\Omega)\, Y^L_M(\Omega)\, d\Omega\, d\psi \tag{3.20b}$$

$$C(L)\, P_T{}^z t^L_M = \langle 2\cos\psi\, Y^L_M(\Omega)\rangle \equiv \int I(\psi;\Omega)\, Y^L_M(\Omega)\, 2\cos\psi\, d\Omega\, d\psi \tag{3.20c}$$

$$C(L)\, P_T{}^x t^L_M = \langle 2\sin\psi\, Y^L_M(\Omega)\rangle \equiv \int I(\psi;\Omega)\, Y^L_M(\Omega)\, 2\sin\psi\, d\Omega\, d\psi \tag{3.20d}$$

with $\Omega = (\theta, \phi)$ and $d\Omega = d(\cos\theta)\, d\phi$. Note that eq. (3.20a) is a particular case of eq. (3.20c) for $L = 0$ (with the conventions ${}^z t_0{}^0 = P_R$, $C(0) = 1/\sqrt{4\pi}$) and is equivalent to eq. (3.16c). By different choices of the initial polarization $\zeta$, i.e. of $P_T$ and $P_L$, one easily deduces from these equations the value of the observables:

i) The target is unpolarized, i.e., $P_L = P_T = 0$. One must first verify that the angular distribution is isotropic around the direction of the beam. Then eq. (3.20b) gives the B-symmetric parameters $t^L_M$, and one must verify that the B-antisymmetric moments vanish.
(We recall that in transversity quantization B-symmetric parameters have $M$ = even and B-antisymmetric parameters have $M$ = odd).

ii) The target is transversally polarized, i.e., $P_T \neq 0$, $P_L = 0$. Eq. (3.20c) gives the B-symmetric polarization transfer parameters ${}^z t^L_M$, and one must verify that the B-antisymmetric moments vanish. Similarly, eq. (3.20d) gives the B-antisymmetric polarization transfer parameters ${}^x t^L_M$ and one must verify that the corresponding B-symmetric moments vanish.
Furthermore one may verify that eq. (3.20b) yields the same results as in case i).

iii) The target is longitudinally polarized, i.e., $P_T = 0$, $P_L \neq 0$. One must verify that the angular distribution is isotropic around the common direction of the beam and of the polarization vector $\zeta$. Then eq. (3.20b) gives the B-symmetric parameters $t^L_M$ (which should be equal to the parameters obtained in i)) and the B-antisymmetric polarization transfer parameters ${}^y t^L_M$.

iv) The target is arbitrarily polarized, i.e., $P_T \neq 0$, $P_L \neq 0$. One obtains all the parameters. Eq. (3.20b) gives the B-symmetric parameters $t^L_M$ and the B-antisymmetric parameters ${}^y t_M{}^L$. Eq. (3.20c, d) give the B-antisymmetric parameters ${}^z t^L_M$ and ${}^x t^L_M$, and one may verify that their B-symmetric moments vanish.
Of course, if the decay is parity conserving, this analysis yields only the $L$ = even parameters (cf. section 2.4.1). The moments with $L$ = odd must be found compatible with zero.

### 3.3.3. Production and joint decay angular distribution

If both final particles (3 and 4) are untable and undergo two body decay (e.g. $\pi p \to \rho\Delta$ $\rho \to \pi\pi$, $\Delta \to N\pi$) the combined production and joint decay angular distribution is (cf. eq. (2.9, 2.19, 3.11))

$$I(\psi;\theta,\phi,\theta',\phi') = \frac{1}{2\pi}\sum_{L=0}^{2j}\sum_{L'=0}^{2j} C(L)\, C'(L') \sum_{M=-L}^{+L}\sum_{M'=-L'}^{+L'} \{\overline{t^{LL'}_{MM'}} + P_T\cos\psi\, \overline{{}^z t^{LL'}_{MM'}}$$

$$+ P_T \sin\psi\, \overline{{}^x t^{LL'}_{MM'}} + P_L \overline{{}^y t^{LL'}_{MM'}}\}\; Y^L_M(\theta,\phi)\; Y^{L'}_{M'}(\theta',\phi') \tag{3.21}$$

where $t_{00}^{00} = 1$, ${}^z t_{00}^{00} = P_R\, C(0) = C'(0) = 1/\sqrt{4\pi}$, $Y_0{}^0 = 1/\sqrt{4\pi}$. The moment analysis of this distribution gives $(\Omega = (\theta, \phi), \Omega' = (\theta', \phi'))$.

$$C(L)\, C'(L')\, (t_{MM'}^{LL'} + P_L{}^y t_{MM'}^{LL'}) = \langle Y_M^L(\Omega)\, Y_{M'}^{L'}(\Omega')\rangle$$

$$C(L)\, C'(L')\, P_T{}^z t_{MM'}^{LL'} = \langle 2 \cos \psi\, Y_M^L(\Omega)\, Y_{M'}^{L'}(\Omega')\rangle \tag{3.22}$$

$$C(L)\, C'(L')\, P_T{}^x t_{MM'}^{LL'} = \langle 2 \sin \psi\, Y_M^L(\Omega)\, Y_{M'}^{L'}(\Omega')\rangle .$$

A similar discussion to that of the preceding section 3.3.2. can be made. We shall not repeat it. We only recall that in the present case, the B-symmetric parameters (in transversity) have $M + M' =$ even and the B-antisymmetric ones have $M + M' =$ odd. Furthermore, if both decays are parity conserving, one gets only the $L =$ even and $L' =$ even parameters; all the moments with $L$ or $L'$ odd must vanish. If the decay in $(\theta', \phi')$ is parity violating one gets the multipole parameters with $L =$ even and $L' =$ even or odd. All other moments vanish.

### 3.3.4. Production and cascade decay angular distribution

Assume again that only one final particle decays, but that it undergoes a cascade decay of the type discussed in section 2.4.3. (e.g. $Kp \to \pi\Sigma^*$, $\Sigma^* \to \Lambda\pi$, $\Lambda \to p\pi$). Then one may study the angular distribution of the production and of the cascade decay. It reads (cf. eq. (2.21a) for the canonical quantization frame and (2.21b) for the helicity quantization frame).

$$I(\psi; \theta, \phi; \theta_1, \phi_1) = \frac{1}{2\pi} \sum_{L=0}^{2j} \sum_{L_1=0}^{1} \sum_{J\,\text{even}} C(L, L_1, J) \sum_{M, M_1, N} \langle JL, NM_1 \mid LM\rangle \{\overline{t_M^L} + P_T \cos \psi \overline{{}^z t_M^L}$$

$$+ P_T \sin \psi \overline{{}^x t_M^L} + P_L \overline{{}^y t_M^L}\}\, Y_N^J(\theta, \phi)\, Y_{M_1}^{L_1}(\theta, \phi_1) \tag{3.23a}$$

$$I(\psi; \theta, \phi; \theta_1{}^h, \phi_1{}^h) = \frac{1}{2\pi} \sum_{L=0}^{2j} \sum_{L_1=0}^{1} C^h(L, L_1, M_1) \sum_{M, M_1} \{\overline{t_M^L} + P_T \cos \psi \overline{{}^z t_M^L}$$

$$+ P_T \sin \psi\, \overline{{}^x t_M^L} + P_L \overline{{}^y t_M^L}\} \sqrt{\frac{2L+1}{4\pi}}\, \overline{D^L(\phi, \theta, 0)_{M_1}^M}\, Y_{M_1}^{L_1}(\theta_1{}^h, \phi_1{}^h). \tag{3.23b}$$

where the coefficients $C(L, L_1, J)$ of Eq. (3.23a) and $C^h(L, L_1, M_1)$ of (3.23b) are given in Eq. (2.22a) and (2.22b). With these known values of the coefficients, the parameters $t_M{}^L$ and ${}^\alpha t_M{}^L (\alpha = x, y, z)$ are deduced by a best fit adjustment of the combined angular distributions or by a moment analysis $(\Omega = (\theta, \phi), \Omega_1 = (\theta_1, \phi_1), \Omega_1{}^h = (\theta_1{}^h, \phi_1{}^h))$

$$C(L, L_1, J)\, (t_M^L + P_L{}^y t_M^L) = \sum_{N, M_1} \langle JL_1 NM_1 \mid LM\rangle \langle Y_N^J(\Omega)\, Y_{M_1}^{L_1}(\Omega_1)\rangle ,$$

$$C(L, L_1, J)\, P_T{}^z t_M^L = \sum_{N, M_1} \langle JL_1 NM_1 \mid LM\rangle \langle 2 \cos \psi\, Y_N^J(\Omega) Y_{M_1}^{L_1}(\Omega_1)\rangle , \tag{3.24a}$$

$$C(L, L_1, J)\, P_T{}^x t_M^L = \sum_{N, M_1} \langle JL_1 NM_1 \mid LM\rangle \langle 2 \sin \psi\, Y_N^J(\Omega) Y_{M_1}^{L_1}(\Omega_1)\rangle ,$$

$$C^h(L, L_1, M_1)\,(t_M^L + P_L{}^y t_M^L) = \sqrt{\frac{2L+1}{4\pi}}\,\langle \overline{D^L(\phi, \theta, 0)_{M_1}^M\; Y_{M_1}^{L_1}(\Omega_1{}^h)}\rangle,$$

$$C^h(L, L_1, M_1)\,P_T{}^z t_M^L = \sqrt{\frac{2L+1}{4\pi}}\,\langle 2\cos\psi \overline{D^L(\phi, \theta, 0)_{M_1}^M\; Y_{M_1}^{L_1}(\Omega_1{}^h)}\rangle, \qquad (3.24\,b)$$

$$C^h(L, L_1, M_1)\,P_T{}^x t_M^L = \sqrt{\frac{2L+1}{4\pi}}\,\langle 2\sin\psi \overline{D^L(\phi, \theta, 0)_{M_1}^M\; Y_{M_1}^{L_1}(\Omega_1{}^h)}\rangle.$$

A discussion identical to that of Section 3.3.2. can be made. Note however that in this case all parameters ($L$ = even and $L$ = odd) can be measured.

### 3.3.5. More complex combined production and decay angular distributions

One may consider more intricate situations. For example, if the two final particles are unstable and one of them undergo a cascade decay (e.g. $\mathrm{Kp} \to \rho\Sigma^*$, $\rho \to \pi\pi$, $\Sigma^* \to \Lambda\pi$, $\Lambda \to \mathrm{p}\pi$) the complete angular distribution involves 7 angles. Still more complex is the case where both final particles undergo cascade decays (e.g. $\mathrm{Kp} \to A_1\Sigma^*$, $A_1 \to \rho\pi$, $\rho \to \pi\pi$, $\Sigma^* \to \Lambda\pi$, $\Lambda \to \mathrm{p}\pi$). Then the complete angular distribution involves 9 angles. The expressions of such angular distributions are easily written down, however the present day experimentalists are not yet interested in such complex reactions with polarized target.

## 4. Amplitude Reconstruction in Usual Reactions

In the previous sections we have shown the way of measuring the observables of a reaction with unpolarized target (section 2) or with polarized target (section 3). They are embodied in the final polarization $\sigma\varrho_f$ or in the transfer polarization matrix $W$, which are quadratic expressions of the transition matrix $T$ or $\tilde{T}$, namely (c.f. eq. (2.14) and (3.9′)).

$$\sigma\varrho_f = \frac{1}{2}\,TT^\dagger, \qquad W = \frac{1}{2}\,\tilde{T}\tilde{T}^\dagger.$$

We call amplitudes the elements of these transition matrices. Their reconstruction consists essentially in obtaining an explicit expression for $T$ or $\tilde{T}$ by inverting the quadratic expressions $TT^\dagger$ or $\tilde{T}\tilde{T}^\dagger$. Theoretically this can easily be done, and one obtains $T$ or $\tilde{T}$ up to some unknown phases (by the procedure of "conventional amplitude reconstruction" of Appendix 1,2). Practically each concrete case needs a separate study since generally the observable matrices $\sigma\varrho_f$ or $W$ are not completely measured.
In this section we present the practical method of reconstructing the amplitudes in usual reactions with unpolarized and/or polarized target. For pedagogical purposes we first recall the method of reconstructing the amplitudes in the simplest reaction type $\pi\mathrm{p} \to \mathrm{K}\Lambda$, by measurement of the classical Wolfenstein parameters $P, R, A$. In view of further generalizations to higher final spins, we introduce, already in this simple case, the multipole formalism and some complex spin rotation parameters. This is discussed in section 4.1., and summarized in table 4. In the following sections we present the details of the generalization to reactions of the types $\pi\mathrm{p} \to \mathrm{K}\Sigma^*$, $\pi\Delta$, $\mathrm{K}^*\Lambda$, $\rho N$ as simple comments to the tables 5 to 10, in which all the recipes for measurements and amplitude reconstructions have been gathered.

Table 4

Amplitude reconstruction for reactions of type $\pi p \to K\Lambda$ with polarized target
(Comparison with the PRA parameters and the spin flip and non flip helicity amplitudes)

---

a) Combined production and decay angular distribution and measurement of the polarization transfer by the method of moments.

$$I(\psi, \theta\phi) = \frac{1}{2\pi} \sum_L C(L) \sum_M \left[\overline{t^L_M} + P_T\left(\cos\psi \overline{t^L_M} + \sin\psi \overline{{}^x t^L_M}\right) + P_L \overline{{}^y t^L_M}\right] Y^L_M(\theta\phi)$$

$$C(L)\,(t^L_M + P_L {}^{l}t^L_M) = \langle Y^L_M(\theta\varphi)\rangle$$

$$C(L)\,P_T {}^z t^L_M = \langle 2\cos\psi\, Y^L_M(\theta\varphi)\rangle$$

$$C(L)\,P_T {}^x t^L_M = \langle 2\sin\psi\, Y^L_M(\theta\varphi)\rangle$$

with: $C(0) = 1/\sqrt{4\pi}$, $C(1) = \alpha_\Lambda/\sqrt{4\pi}$ (all other $C$ coefficients vanish)

---

b) The 8 real observables in transversity quantization

$$\sigma_0,\ t_0{}^1$$

$$P = {}^z t_0{}^0,\ {}^z t_0{}^1,\ \mathrm{Re}\,{}^x t_1{}^1,\ \mathrm{Im}\,{}^x t_1{}^1$$

$$\mathrm{Re}\,{}^y t_1{}^1,\ \mathrm{Im}\,{}^y t_1{}^1$$

($M$ = even for $t^L_M$, ${}^z t^L_M$, and odd for ${}^x t^L_M$, ${}^y t^L_M$ by B-symmetry; $t^L_{-M} = (-)^M \overline{t^L_M}$)

---

c) The 4 linear constraints and the 3 spin rotation parameters

| | |
|---|---|
| $P_0 = \frac{1}{2}\left[1 + \sqrt{3}\, t_0{}^1\right] = \frac{1}{2}\left[P + \sqrt{3}\,{}^z t_0{}^1\right] = P_0$ | $P_0 = \frac{1}{2}(1 + P)$ |
| $P_0' = \frac{1}{2}\left[1 - \sqrt{3}\, t_0{}^1\right] = \frac{-1}{2}\left[P - \sqrt{3}\,{}^z t_0{}^1\right] = P_0'$ | $P_0' = \frac{1}{2}(1 - P)$ |
| $P_0 + P_0' = 1^\dagger$ | |
| $R_0 = \frac{1}{2}\left[-\sqrt{6}\,\overline{{}^y t_1{}^1}\right] = \frac{1}{2}\left[-\sqrt{6}\,\overline{{}^x t_1{}^1}\right] = R_0$ | $R_0 = \frac{1}{2}(R + iA)$ |

---

d) The non linear constraint

| | |
|---|---|
| $\lvert R_0\rvert^2 = P_0 P_0'$ $\quad P_0 \geqq 0$ | $P^2 + R^2 + A^2 = 1$ |

---

† The two first constraints imply $\sqrt{3}\, t_0{}^1 = P$, $\sqrt{3}\,{}^z t_0{}^1 = 1$

Table 4 — cont'd

e) Transfer of polarization

| | |
|---|---|
| | $\sigma = \sigma_0(1 + Pz)$ |
| $\sigma\varrho_{++} = \sigma_0 P_0(1 + z)$ | $\sigma Z = \sigma_0(P + z)$ |
| $\sigma\varrho_{--} = \sigma_0 P_0'(1 - z)$ | $\sigma X = \sigma_0(Rx + Ay)$ |
| $\sigma\varrho_{+-} = \sigma_0 R_0(x - iy)$ | $\sigma Y = \sigma_0(-Ax + Ry)$ |

f) Transversity amplitudes and the usual spin non flip and spin flip helicity amplitudes

$$\begin{array}{c|cc} & \multicolumn{2}{c}{\lambda_p} \\ \lambda_\Lambda & +\frac{1}{2} & -\frac{1}{2} \\ \hline +\frac{1}{2} & a & \\ -\frac{1}{2} & & a' \end{array} = T_{2\lambda_\Lambda, 2\lambda_p} \qquad \begin{array}{|cc|} F & G \\ -G & F \end{array} = H_{2\lambda_\Lambda, 2\lambda_p}$$

$$a = F + iG$$

$$a' = F - iG$$

f') Collinearity constraints

$$G = 0 \Leftrightarrow a = a'$$

g) Expression of the observables of c), e) as functions of the transversity amplitudes

$$\sigma_0 = \frac{1}{2}(|a|^2 + |a'|^2) = |F|^2 + |G|^2$$

| | | |
|---|---|---|
| $2\sigma_0 P_0 = \lvert a\rvert^2$ | $\sigma_0 P = \frac{1}{2}(\lvert a\rvert^2 - \lvert a'\rvert^2) = 2 \operatorname{Im} F\bar{G}$ | |
| $2\sigma_0 P_0' = \lvert a'\rvert^2$ | $\sigma_0 R = \operatorname{Re} a\overline{a'}$ | $= \lvert F\rvert^2 - \lvert G\rvert^2$ |
| $2\sigma_0 R_0 = a\bar{a}'$ | $\sigma_0 A = \operatorname{Im} a\overline{a'}$ | $= 2 \operatorname{Re} F\bar{G}$ |

Table 5

Amplitude reconstruction for reactions of type $\pi p \to K\Sigma^*(0^-1/2^+ \to 0^-3/2^+)$ with unpolarized target

a) Angular distribution of the $\Sigma^*$ decay and measurement of the even multipole parameters by the method of moments.

$$I(\theta, \phi) = \sum_L C(L) \sum_M \overline{t^L_M} Y^L_M(\theta, \phi)$$

$$C(L)\, t^L_M = \langle Y^L_M(\theta, \phi)\rangle$$

with: $C(0) = 1/\sqrt{4}\,\pi$, $\quad C(2) = -1/\sqrt{4}\,\pi$ (all other $C$ coefficients vanish)

Table 5 — cont'd

b) Cascade angular distribution of $\Sigma^*$ decay ($\Sigma^* \to \Lambda\pi$, $\Lambda \to p\pi$) and measurement of all multipole parameters by the method of moments.

b 1) Canonical frame for the $\Lambda$

$$I(\theta, \phi, \theta_1, \phi_1) = \sum_{LL_1J} C(LL_1J) \sum_{MNM_1} \langle JL_1NM_1 \mid LM \rangle \overline{t^L_M} Y^J_N(\theta\phi)\, Y^{L_1}_{M_1}(\theta_1\phi_1)$$

$$C(LL_1J)\, t^L_M = \sum_N \langle JL_1NM_1 \mid LM \rangle \langle Y^J_N(\theta, \phi)\, Y^{L_1}_{M_1}(\theta_1, \phi_1) \rangle$$

b 2) Helicity frame for the $\Lambda$

$$I(\theta, \phi, \theta_1{}^h, \phi_1{}^h) = \sum_{LL_1M_1} C^h(LL_1M_1) \sum_M \overline{t^L_M} \sqrt{\frac{2L+1}{4\pi}}\, \overline{D^L(\phi\theta 0)^{M_1}_M}\, Y^{L_1}_{M_1}(\theta_1{}^h, \phi_1{}^h)$$

$$C^h(LL_1M_1)\, t^L_M = \sqrt{\frac{2L+1}{4\pi}} \langle \overline{D^L(\phi\theta 0)^M_{M_1}}\, Y^{L_1}_{M_1}(\theta_1{}^h, \phi_1{}^h) \rangle$$

b 3) Values of the non vanishing coefficients in b 1) and b 2)

| | |
|---|---|
| $C(000) = 1/4\pi,$ | $C^h(000) = 1/4\pi,$ |
| $C(202) = -1/4\pi,$ | $C^h(200) = -1/4\pi,$ |
| $C(110) = (\alpha/4\pi)\sqrt{5/9},$ | $C^h(110) = (\alpha/4\pi)\sqrt{1/15},$ |
| $C(112) = (\alpha/4\pi)\sqrt{2/45},$ | $C^h(11\pm 1) = (\alpha/4\pi)\sqrt{4/15}$ |
| $C(312) = -(\alpha/4\pi)\sqrt{7/5},$ | $C^h(310) = -(\alpha/4\pi)\sqrt{3/5}$ |
| | $C^h(31\pm 1) = -(\alpha/4\pi)\sqrt{2/5}$ |

c) The 8 real observables in transversity quantization

$\sigma_0$, $t_0{}^2$, Re $t_2{}^2$, Im $t_0{}^1$, $t_0{}^3$, Re $t_2{}^3$, Im $t_2{}^3$.

($M$ = even by B-symmetry, $t^L_{-M} = (-1)^M t_M{}^L$)

d) Observable density matrix elements in transversity quantization

$$P_1 \equiv \varrho_{11} = 1/4 \left[1 - \sqrt{5}\, t_0{}^2 + \sqrt{3/5}\, t_0{}^1 - \sqrt{63/5}\, t_0{}^3\right]$$

$$P_1' \equiv \varrho_{-1-1} = 1/4 \left[1 - \sqrt{5}\, t_0{}^2 - \sqrt{3/5}\, t_0{}^1 + \sqrt{63/5}\, t_0{}^3\right]$$

$$P_2 \equiv \varrho_{-3-3} = 1/4 \left[1 + \sqrt{5}\, t_0{}^2 - \sqrt{27/5}\, t_0{}^1 - \sqrt{7/5}\, t_0{}^3\right]$$

$$P_2' \equiv \varrho_{33} = 1/4 \left[1 + \sqrt{5}\, t_0{}^2 + \sqrt{27/5}\, t_0{}^1 + \sqrt{7/5}\, t_0{}^3\right]$$

$$Q \equiv \varrho_{1-3} = 1/4 \left[\sqrt{10}\, \overline{t_2{}^2} - \sqrt{14}\, \overline{t_2{}^3}\right], \qquad Q' \equiv \varrho_{3-1} = 1/4 \left[\sqrt{10}\, \overline{t_2{}^2} + \sqrt{14}\, \overline{t_2{}^3}\right]$$

e) Positivity and rank constraints

$$P_1 \geqq 0, \quad P_1' \geqq 0, \quad P_2 \geqq 0, \quad P_2' \geqq 0, \quad |Q|^2 = P_1 P_2, \quad |Q'^2| = P_1' P_2'$$

Table 5 — cont'd

---

f) Transversity and helicity amplitudes

$$\begin{array}{c|cc|}
 & \multicolumn{2}{c}{\lambda_p} \\
\lambda_{\Sigma^*} & \frac{1}{2} & -\frac{1}{2} \\
\hline
3/2 & . & b' \\
1/2 & a & . \\
-1/2 & . & a' \\
-3/2 & b & . \\
\hline
\end{array} = T_{2\lambda_{\Sigma^*},2\lambda_p} \qquad \begin{array}{|cc|}
\hline
B' & -B \\
A & A' \\
A' & -A \\
B & B' \\
\hline
\end{array} = H_{2\lambda_{\Sigma^*},2\lambda_p}$$

$$A + iA' = -1/2\left(a + \sqrt{3}\,b\right) \qquad B + iB' = -1/2\left(\sqrt{3}\,a - b\right)$$

$$A - iA' = 1/2\left(a' + \sqrt{3}\,b'\right) \qquad B - iB' = 1/2\left(\sqrt{3}\,a' - b'\right)$$

f') Collinearity constraints

$$A' = B' = B = 0 \Leftrightarrow a = -a' = \sqrt{\frac{1}{3}}\,b = -\sqrt{\frac{1}{3}}\,b'$$

---

g) Expression of the observables in d) and e) as functions of the transversity amplitudes

$$2\sigma_0 P_1 = |a|^2 \qquad 2\sigma_0 P_1' = |a'|^2$$

$$2\sigma_0 P_2 = |b|^2 \qquad 2\sigma_0 P_2' = |b'|^2$$

$$2\sigma_0 Q = a\bar{b} \qquad 2\sigma_0 Q' = b'\bar{a}'$$

Table 6

Amplitude reconstruction for reactions of type $p\pi \to K\Sigma^*$ ($0^- \, 1/2^+ \to 0^- \, 3/2^+$) with polarized target.

---

a) Combined production and cascade angular distribution for general target polarization, and measurement of the polarization transfer by the method of moments

a 1) Canonical frame for the $\Lambda$

$$I(\psi, \theta, \Phi, \theta_1, \Phi_1) = \frac{1}{2\pi} \sum_{LJL_1} C(LL_1J) \sum_{MNM_1} \langle JL_2NM_1 \,|LM\rangle \left[t^L_M + P_T\left(\cos\psi\,\overline{{}^z t^L_M} + \sin\psi\,\overline{{}^x t^L_M}\right) + P_L\overline{{}^y t^L_M}\right] Y^J_N(\theta, \Phi)\, Y^L_M(\theta_1, \Phi_1)$$

$$C(LL_1J)\,(t^L_M + P_L{}^y t^L_M = \sum_{NM_1} \langle JL_1NM_1 \mid LM\rangle \langle Y^J_N(\theta, \Phi)\, Y^{L_1}_{M_1}(\theta_1, \Phi_1)\rangle$$

$$C(LL_1J)\, P_T{}^z t^L_M = \sum_{NM_1} \langle JL_1NM_1 \mid LM\rangle \langle 2\cos\psi\, Y^J_N(\theta, \Phi)\, Y^{L_1}_{M_1}(\theta_1, \Phi_1)$$

$$C(LL_1J)\, P_T{}^x t^L_M = \sum_{NM_1} \langle JL_1NM_1 \mid LM\rangle \langle 2\sin\psi\, Y^J_N(\theta, \Phi)\, Y^{L_1}_{M_1}(\theta_1, \Phi_1)$$

Table 6 — cont'd

a 2) Helicity frame for the $\Lambda$

$$I(\psi, \theta, \phi, \theta_1^h, \phi_1^h) = \frac{1}{2\pi} \sum_{LL_1M_1} C^h(LL_1 M_1) \sum_M \Big[\overline{t^L_M} + P_T\big(\cos\psi\, \overline{{}^z t^L_M} + \sin\psi\, \overline{{}^x t^L_M}\big) + P_L\, \overline{{}^y t^L_M}\Big] \sqrt{\frac{2L+1}{4\pi}}\, \overline{D^L(\phi\theta 0)^M_{M_1}}\, Y^{L_1}_{M_1}(\theta_1^h, \phi_1^h)$$

$$C^h(LL_1M_1)\,(t^L_M + P_L{}^y t^L_M) = \sqrt{\frac{2L+1}{4\pi}} \langle \overline{D^L(\phi\theta 0)^M_{M_1}}\, Y^{L_1}_{M_1}(\theta_1^h, \phi_1^h)\rangle$$

$$C^h(LL_1M_1)\, P_T{}^z t^L_M = \sqrt{\frac{2L+1}{4\pi}} \langle 2\cos\psi\, \overline{D^L(\phi\theta 0)^M_{M_1}}\, Y^{L_1}_{M_1}(\theta_1^h, \phi_1^h)\rangle$$

$$C^h(LL_1M_1)\, P_T{}^x t^L_M = \sqrt{\frac{2L+1}{4\pi}} \langle 2\sin\psi\, \overline{D^L(\phi\theta 0)^M_{M_1}}\, Y^{L_1}_{M_1}(\theta_1^h, \phi_1^h)\rangle$$

with the coefficients $C(LL_1J)$ and $C^h(LL_1M_1)$ of Table 5 b 3)

---

b) The 32 real observables in transversity quantization

$\sigma_0$, $t_0^2$, Re $t_2^2$, Im $t_2^2$, $t_0^1$, $t_0^3$, Re $t_2^3$, Im $t_2^3$,

$P_R = {}^z t_0^0$, ${}^z t_0^2$, Re ${}^z t_2^2$, Im ${}^z t_2^2$, ${}^z t_0^1$, ${}^z t_0^3$, Re ${}^z t_2^3$, Im ${}^z t_2^3$,

Re ${}^x t_1^2$, Im ${}^x t_1^2$, Re ${}^x t_1^1$, Im ${}^x t_1^1$, Re ${}^x t_1^3$, Im ${}^x t_1^3$, Re ${}^x t_3^3$, Im ${}^x t_3^3$,

Re ${}^y t_1^2$, Im ${}^y t_1^2$, Re ${}^y t_1^1$, Im ${}^y t_1^1$, Re ${}^y t_1^3$, Im ${}^y t_1^3$, Re ${}^y t_3^3$, Im ${}^y t_3^3$.

$\big(M =$ even for $t^L_M$ and ${}^z t^L_M$, $M =$ odd for ${}^x t^L_M$ and ${}^y t^L_M$ by B-symmetry, $t^L_{-M} = (-1)^M\, \overline{t^L_M}\big)$

---

c) The 16 linear constraints and the 15 generalized spin rotation parameters

$$P_1 = \frac{1}{4}\left[1 - \sqrt{5}\, t_0^2 + \sqrt{\frac{3}{5}}\, t_0^1 - \sqrt{\frac{63}{5}}\, t_0^3\right] = \frac{1}{4}\left[P_R - \sqrt{5}\, {}^z t_0^2 + \sqrt{\frac{3}{5}}\, {}^z t_0^1 - \sqrt{\frac{63}{5}}\, {}^z t_0^3\right] = P_1$$

$$P_1' = \frac{1}{4}\left[1 - \sqrt{5}\, t_0^2 - \sqrt{\frac{3}{5}}\, t_0^1 + \sqrt{\frac{63}{5}}\, t_0^3\right] = \frac{-1}{4}\left[P_R - \sqrt{5}\, {}^z t_0^2 - \sqrt{\frac{3}{5}}\, {}^z t_0^1 + \sqrt{\frac{63}{5}}\, {}^z t_0^3\right] = P_1'$$

$$P_2 = \frac{1}{4}\left[1 + \sqrt{5}\, t_0^2 - \sqrt{\frac{27}{5}}\, t_0^1 - \sqrt{\frac{7}{5}}\, t_0^3\right] = \frac{1}{4}\left[P_R + \sqrt{5}\, {}^z t_0^2 - \sqrt{\frac{27}{5}}\, {}^z t_0^1 - \sqrt{\frac{7}{5}}\, {}^z t_0^3\right] = P_2$$

$$P_2' = \frac{1}{4}\left[1 + \sqrt{5}\, t_0^2 + \sqrt{\frac{27}{5}}\, t_0^1 + \sqrt{\frac{7}{5}}\, t_0^3\right] = \frac{-1}{4}\left[P_R + \sqrt{5}\, {}^z t_0^2 + \sqrt{\frac{27}{5}}\, {}^z t_0^1 + \sqrt{\frac{7}{5}}\, {}^z t_0^3\right] = P_2'$$

$$P_1 + P_1' + P_2 + P_2' = 1, \qquad P_1 - P_1' + P_2 - P_2' = P_R$$

Table 6 — cont'd

c) (continued)

$$Q = \frac{1}{4}\left[\sqrt{10}\,\overline{t_2^{\,2}} - \sqrt{14}\,\overline{t_2^{\,3}}\right] = \frac{1}{4}\left[\sqrt{10}\,\overline{{}^{z}t_2^{\,2}} - \sqrt{14}\,\overline{{}^{z}t_2^{\,3}}\right] = Q$$

$$Q' = \frac{1}{4}\left[\sqrt{10}\,\overline{t_2^{\,2}} + \sqrt{14}\,\overline{t_2^{\,3}}\right] = \frac{1}{4}\left[\sqrt{10}\,\overline{{}^{z}t_2^{\,2}} + \sqrt{14}\,\overline{{}^{z}t_2^{\,3}}\right] = Q'$$

$$R = \frac{i}{4}\left[-\sqrt{10}\,\overline{{}^{y}t_1^{\,2}} - \sqrt{\frac{18}{5}}\,\overline{{}^{y}t_1^{\,1}} - \sqrt{\frac{28}{5}}\,\overline{{}^{y}t_1^{\,3}}\right] = \frac{1}{4}\left[-\sqrt{10}\,\overline{{}^{x}t_1^{\,2}} - \sqrt{\frac{18}{5}}\,\overline{{}^{x}t_1^{\,1}} - \sqrt{\frac{28}{5}}\,\overline{{}^{x}t_1^{\,3}}\right] = R$$

$$R' = \frac{i}{4}\left[\sqrt{10}\,\overline{{}^{y}t_1^{\,2}} - \sqrt{\frac{18}{5}}\,\overline{{}^{y}t_1^{\,1}} - \sqrt{\frac{28}{5}}\,\overline{{}^{y}t_1^{\,3}}\right] = \frac{1}{4}\left[\sqrt{10}\,\overline{{}^{x}t_1^{\,2}} - \sqrt{\frac{18}{5}}\,\overline{{}^{x}t_1^{\,1}} - \sqrt{\frac{28}{5}}\,\overline{{}^{x}t_1^{\,3}}\right] = R'$$

$$R_1 = \frac{i}{4}\left[-\sqrt{\frac{24}{5}}\,\overline{{}^{y}t_1^{\,1}} - \sqrt{\frac{84}{5}}\,\overline{{}^{y}t_1^{\,3}}\right] = \frac{1}{4}\left[-\sqrt{\frac{24}{5}}\,\overline{{}^{x}t_1^{\,1}} - \sqrt{\frac{84}{5}}\,\overline{{}^{x}t_1^{\,3}}\right] = R_1$$

$$R_2 = \frac{-i}{4}\left[-\sqrt{28}\,\overline{{}^{y}t_3^{\,3}}\right] = \frac{1}{4}\left[-\sqrt{28}\,\overline{{}^{x}t_3^{\,3}}\right] = R_2$$

d) The 9 non linear constraints and the positivity conditions

$$|Q|^2 = P_1P_2,\quad |R_1|^2 = P_1P_1',\quad P_1R' = Q\bar{R}_1,\quad P_1 \geqq 0,\quad P_2 \geqq 0$$

$$|Q'|^2 = P_1'P_2',\quad RR' = \bar{R}_1R_2,\quad P_1'R = Q'\bar{R}_1,\quad P_1' \geqq 0,\quad P_2' \geqq 0$$

e) Transfer of polarization

| | |
|---|---|
| $\sigma\varrho_{11} = \sigma_0P_1(1+z)$ | $\sigma\varrho_{-1-1} = \sigma_0P_1'(1-z)$ |
| $\sigma\varrho_{-3-3} = \sigma_0P_2(1+z)$ | $\sigma\varrho_{33} = \sigma_0P_2'(1-z)$ |
| $\sigma\varrho_{1-3} = \sigma_0Q(1+z)$ | $\sigma\varrho_{3-1} = \sigma_0Q'(1-z)$ |
| $\sigma\varrho_{31} = \sigma_0R(x+iy)$ | $\sigma\varrho_{-1-3} = \sigma_0R'(x+iy)$ |
| $\sigma\varrho_{1-1} = \sigma R_1(x-iy)$ | $\sigma\varrho_{3-3} = \sigma_0R_2(x+iy)$ |

f) Transversity and helicity amplitudes as in Table 5f)

g) Expression of the observables in c) and e) as functions of the transversity amplitudes)

| | |
|---|---|
| $2\sigma_0P_1 = \lvert a\rvert^2$ | $2\sigma_0P_1' = \lvert a'\rvert^2$ |
| $2\sigma_0P_2 = \lvert b\rvert^2$ | $2\sigma_0P_2' = \lvert b'\rvert^2$ |
| $2\sigma_0Q = a\bar{b}$ | $2\sigma_0Q' = b'\bar{a}'$ |
| $2\sigma_0R = b'\bar{a}$ | $2\sigma_0R' = a'\bar{b}$ |
| $2\sigma_0R_1 = a\bar{a}'$ | $2\sigma_0R_2 = b'\bar{b}$ |

Table 7

Amplitude reconstruction for reactions of type $\pi p \to \pi\Delta(0^-1/2^+ \to 0^-3/2^+)$ with polarized target

---

a) Combined production and decay angular distribution and measurement of the polarization transfer by the method of moments

$$I(\psi, \theta, \phi) = \frac{1}{2\pi} \sum_L C(L) \sum_M \left[\overline{t^L_M} + P_T(\cos\psi \overline{{}^z t^L_M} + \sin\psi \overline{{}^x t^L_M}) + P_L \overline{{}^y t^L_M}\right] Y^L_M(\theta, \phi)$$

$$C(L)\,(t^L_M + P_L {}^y t^L_M) = \langle Y^L_M(\theta, \phi)\rangle$$

$$C(L)\,P_T {}^z t^L_M = \langle 2\cos\psi\, Y^L_M(\theta, \phi)\rangle$$

$$C(L)\,P_T {}^x t^L_M = \langle 2\sin\psi\, Y^L_M(\theta, \phi)\rangle$$

with $C(0) = 1/\sqrt{4\pi}$, $C(2) = 1/\sqrt{4\pi}$ (all other $C$ coefficients vanish)

---

b) The 12 real observables in transversity quantization

$\sigma_0$, $t_0^2$, Re $t_2^2$, Im $t_2^2$

$P_R = {}^z t_0^0$, ${}^z t_0^2$, Re ${}^z t_2^2$, Im ${}^x t_2^2$, Re ${}^x t_1^2$, Im ${}^x t_1^2$, Re ${}^y t_1^2$, Im ${}^y t_1^2$.

($M$ = even for $t^L_M$ and ${}^z t^L_M$, $M$ = odd for ${}^x t^L_M$ and ${}^y t^L_M$, by B-symmetry; $t^L_{-M} = (-1)^M\, t^L_M$)

---

c) The 9 generalized spin rotation parameters and the 2 linear constraints

$$P_1 = \frac{1}{4}\left[1 + P_R - \sqrt{5}\,(t_0^2 + {}^z t_0^2)\right]$$

$$P_1' = \frac{1}{4}\left[1 - P_R - \sqrt{5}\,(t_0^2 - {}^z t_0^2)\right] \qquad P_1 + P_1' + P_2 + P_2' = 1$$

$$P_2 = \frac{1}{4}\left[1 + P_R + \sqrt{5}\,(t_0^2 + {}^z t_0^2)\right] \qquad P_1 - P_1' + P_2 - P_2' = P_R$$

$$P_2' = \frac{1}{4}\left[1 - P_R + \sqrt{5}\,(t_0^2 - {}^z t_0^2)\right]$$

$$Q = \frac{1}{4}\sqrt{10}\,\left(\overline{t_0^2} + \overline{{}^z t_2^2}\right), \qquad Q' = \frac{1}{4}\sqrt{10}\,\left(\overline{t_2^2} - \overline{{}^z t_2^2}\right)$$

$$R_- = -\frac{1}{2}\sqrt{10}\,\overline{{}^x t_1^2} = \frac{i}{2}\sqrt{10}\,\overline{{}^y t_1^2}$$

---

d) The 3 non linear constraints and the positivity conditions

$$|Q|^2 = P_1 P_2, \quad |Q'|^2 = P_1' P_2', \quad |R_-|^2 + |Q + Q'|^2 = (P_1 + P_1')\,(P_2 + P_2')$$

$$P_1 \geqq 0, \quad P_2 \geqq 0, \quad P_1' \geqq 0, \quad P_2' \geqq 0.$$

Table 7 — cont'd

---

e) Transfer of polarization

$$\sigma\varrho^e_{11} = \sigma_0 \left[\frac{1}{2}(P_1 + P_1') + \frac{1}{2}(P_1 - P_1')\, z\right]$$

$$\sigma\varrho^e_{33} = \sigma_0 \left[\frac{1}{2}(P_2 + P_2') + \frac{1}{2}(P_2 - P_2')\, z\right]$$

$$\sigma\varrho^e_{3-1} = \sigma_0 \left[\frac{1}{2}(Q + Q') + \frac{1}{2}(Q - Q')\, z\right]$$

$$\sigma\varrho^e_{31} = \sigma_0 R_-(x + iy)$$

---

f) Transversity and helicity amplitudes

$$\begin{array}{c|cc|}
 & \multicolumn{2}{c}{\lambda_p} \\
\lambda_\Delta & \frac{1}{2} & -\frac{1}{2} \\
\hline
3/2 & \cdot & b' \\
1/2 & a & \cdot \\
-1/2 & \cdot & a' \\
3/2 & b & \cdot \\
\hline
\end{array} = T_{2\lambda_\Delta, 2\lambda_p} \qquad
\begin{array}{|cc|}
\hline
B' & -B \\
A & A' \\
A' & -A \\
B & B' \\
\hline
\end{array} = H_{2\lambda_\Delta, 2\lambda_p}$$

$$A + iA' = -\frac{1}{2}\left(a + \sqrt{3}\, b\right) \qquad B + iB' = -\frac{1}{2}\left(\sqrt{3}\, a - b\right)$$

$$A - iA' = \frac{1}{2}\left(a' + \sqrt{3}\, b'\right) \qquad B - iB' = \frac{1}{2}\left(\sqrt{3}\, a' - b'\right)$$

f') Collinearity constraints

$$A' = B' = B = 0 \Leftrightarrow a = -a' = \sqrt{\frac{1}{3}}\, b = -\sqrt{\frac{1}{3}}\, b'$$

---

g) Expression of the observables in c)—e) in terms of the transversity amplitudes in f)

$$2\sigma_0 P_1 = |a|^2 \qquad 2\sigma_0 P_1' = |a'|^2$$

$$2\sigma_0 P_2 = |b|^2 \qquad 2\sigma_0 P_2' = |b'|^2$$

$$2\sigma_0 Q = a\bar{b} \qquad 2\sigma_0 Q' = b'\bar{a}'$$

$$2\sigma_0 R_- = b'\bar{a} - a'\bar{b} \rightarrow [a\bar{a}' = 2\sigma_0(P_1 Q' - P_1' Q)/R_-]$$

Table 8

Amplitude reconstruction for reactions of type $\pi p \to K^*\Lambda(0^- 1/2^+ \to 1^{-e}\, 1/2^+$ with unpolraized target

---

a) Joint angular distribution of K* and Λ decays and measurement of the double multipole parameters by the method of moments

$$I(\theta, \phi, \theta', \phi') = \sum_{L,L'} C_{K^*}(L)\, C_\Lambda(L') \sum_{M,M'} \overline{t^{LL'}_{MM'}}\, Y^L_M(\theta, \phi)\; Y^{L'}_{M'}(\theta, \phi)$$

$$C_{K^*}(L)\; C_\Lambda(L')\; t^{LL'}_{MM'} = \langle Y^L_M(\theta, \phi) Y^{L'}_{M'}(\theta'\; \phi')\rangle$$

with

$$C_{K^*}(0) = C_\Lambda(0) = 1/\sqrt{4\pi}, \qquad C_{K^*}(2) = -1/\sqrt{2\pi}, \quad C_\Lambda(1) = \alpha_\Lambda/\sqrt{4\pi}$$

(all other $C$ coefficients vanish)

---

b) The 12 real observables in transversity quantization

$$\sigma_0,\, t^{20}_{00},\, t^{01}_{00},\, t^{21}_{00}, \quad \left.\begin{matrix}\mathrm{Re}\\ \mathrm{Im}\end{matrix}\right\}\; t^{20}_{20},\, t^{21}_{20},\, t^{21}_{11},\, t^{21}_{1-1}$$

$$\left(M + M' = \text{even, by B-symmetry, } t^{LL'}_{-M-M'} = (-1)^{M+M'}\, \overline{t^{LL'}_{MM'}}\right)$$

---

c) Observable density matrix elements in transversity quantization

$$\left.\begin{matrix} P_0 = \varrho^{00}_{++} \\ P_0{}' = \varrho^{00}_{--}\end{matrix}\right\} = \frac{1}{6}\left[1 - \sqrt{10}\, t^{20}_{00} \pm \left(\sqrt{3}\, t^{01}_{00} - \sqrt{30}\, t^{21}_{00}\right)\right]$$

$$\left.\begin{matrix} P = \varrho^{11e}_{--} \\ P' = \varrho^{11e}_{++}\end{matrix}\right\} = \frac{1}{6}\left[1 + \sqrt{\frac{5}{2}}\; t^{20}_{00} \mp \left(\sqrt{3}\, t^{01}_{00} + \sqrt{\frac{15}{2}}\; t^{21}_{00}\right)\right]$$

$$\left.\begin{matrix} Q = \varrho^{1-1}_{--} \\ Q' = \varrho^{1-1}_{++}\end{matrix}\right\} = \frac{1}{6}\left[\sqrt{15}\, \overline{t^{20}_{20}} \mp \sqrt{45}\, \overline{t^{21}_{20}}\right]$$

$$S_1 = \varrho^{10e}_{-+} = \frac{-1}{6}\sqrt{45}\, \overline{t^{21}_{1-1}}$$

$$S_2 = \varrho^{10e}_{+-} = \frac{1}{6}\sqrt{45}\, \overline{t^{21}_{11}}$$

---

d) Positivity and diacritical constraints

$$P_0 \geqq 0, \qquad P \geqq |Q|, \qquad P_0{}' \geqq 0, \qquad P' \geqq |Q'|$$

$$32 P_0 P = |\Gamma - \sqrt{\Delta}|^2/|S_2|^2 + |\Gamma + \sqrt{\Delta}|^2/|S_1|^2$$

$$32 P_0{}' P' = |\Gamma' + \sqrt{\Delta}|^2/|S_2|^2 + |\Gamma' - \sqrt{\Delta}|^2/|S_1|^2$$

Table 8 — cont'd

with

$\Gamma = 4S_1S_2 - P_0Q + P_0'Q'$ $\qquad$ $\Gamma' = 4S_1S_2 + P_0Q - P_0'Q'$

$\Delta = \Delta(4S_1S_2, -P_0Q, -P_0'Q']$ $\qquad$ $\Delta(x, y, z) = x^2 + y^2 + z^2 - 2(xy + yz + zx)$

$\sqrt{\Delta}$ = one of the two complex roots, to be fixed by the diacritical constraints.

e) Transversity and helicity amplitudes

| $\lambda_\Lambda$ | $\lambda_{K^*}$ \ $\lambda_p$ | $+\frac{1}{2}$ | $-\frac{1}{2}$ |
|---|---|---|---|
| $+\frac{1}{2}$ | 1 | . | $c'$ |
| | 0 | $a$ | . |
| | $-1$ | . | $b'$ |
| $-\frac{1}{2}$ | 1 | $b$ | . |
| | 0 | . | $a'$ |
| | $-1$ | $c$ | . |

$= T^{\lambda_{K^*}}_{2\lambda_\Lambda, 2\lambda_p}$;

| | |
|---|---|
| $C'$ | $-C$ |
| $A$ | $A'$ |
| $B'$ | $-B$ |
| $B$ | $B'$ |
| $A'$ | $-A'$ |
| $C$ | $C$ |

$= H^{\lambda_{K^*}}_{2\lambda_\Lambda, 2\lambda_p}$

$2(A + iA') = -\sqrt{2}\,(b + c),$ $\qquad$ $2(A - iA') = \sqrt{2}\,(b' + c')$

$2(B + iB') = -\sqrt{2}\,a + (b - c),$ $\qquad$ $2(B - iB') = \sqrt{2a'} - (b' - c'),$

$2(C + iC') = -\sqrt{2}\,a - (b - c),$ $\qquad$ $2(C - iC') = \sqrt{2}\,a' + (b' - c').$

The BYERS-YANG [5] amplitudes are given by

$a_+ = -a', \quad b_+ = -(b' + c')/\sqrt{2},$ $\qquad$ $c_+ = -i(-b' + c')/\sqrt{2}$

$a_- = -a, \quad b_- = (b + c)/\sqrt{2},$ $\qquad$ $c_- = i(-b + c)/\sqrt{2}$

if one uses standard $s$, $t$, and $u$ transversity frames for the quantization of p, K*, and $\Lambda$ respectively.

e') Collinearity constraints

$$A' = B' = C' = C = 0 \Leftrightarrow b = -b', \quad a = -a' = \sqrt{\frac{1}{2}}\,(c - b), \quad c = -c'.$$

f) Expression of the observables in c) and d) as functions of the transversity amplitudes

$2\sigma P_0 = |a|^2$ $\qquad$ $2\sigma P_0' = |a'|^2$

$2\sigma P = \frac{1}{2}\,(|b|^2 + |c|^2)$ $\qquad$ $2\sigma P' = \frac{1}{2}\,(|b'|^2 + |c'|^2)$

$2\sigma Q = b\bar{c}$ $\qquad$ $2\sigma Q' = c'\bar{b}'$

$2\sigma S_1 = \frac{1}{2}\,(b\bar{a} - a'\bar{b}')$ $\qquad$ $2\sigma S_2 = \frac{1}{2}\,(c'\bar{a}' - a\bar{c})$

Table 8 — cont'd

g) Algebraic reconstruction of the amplitudes

$$|a|^2 = 2\sigma P_0 \qquad |a'|^2 = 2\sigma P_0$$

$$|b|^2 = 2\sigma\left(P + \varepsilon\sqrt{P^2 - |Q|^2}\right) \qquad |c'|^2 = 2\sigma\left(P' + \varepsilon\sqrt{P'^2 - |Q'|^2}\right)$$

$$|c|^2 = 2\sigma\left(P - \varepsilon\sqrt{P^2 - |Q|^2}\right) \qquad |c'|^2 = 2\sigma\left(P' - \varepsilon\sqrt{P'^2 - |Q'|^2}\right)$$

$$\phi_b - \phi_c = \operatorname{Arg} Q \qquad \phi_{c'} - \phi_{b'} = \operatorname{Arg} Q'$$

$$\begin{cases}\phi_b - \phi_a = \operatorname{Arg}\left(\Gamma - \sqrt{\Delta}\right)/S_2 \\ \phi_a - \phi_c = \operatorname{Arg}\left(-\Gamma - \sqrt{\Delta}\right)/S_1\end{cases} \qquad \begin{cases}\phi_{a'} - \phi_{b'} = \operatorname{Arg}\left(-\Gamma' - \sqrt{\Delta}\right)/S_2 \\ \phi_{c'} - \phi_{b'} = \operatorname{Arg}\left(\Gamma' - \sqrt{\Delta}\right)/S_1\end{cases}$$

with the expression for $\Gamma$, $\Gamma'$, $\sqrt{\Delta}$ given in d), and the signs $\varepsilon$, $\varepsilon'$ fixed by:

$$\varepsilon[|S_1|^2\,|\Gamma - \sqrt{\Delta}|^2 - |S_2|^2\,|\Gamma + \sqrt{\Delta}|^2] \gtreqq 0, \qquad \varepsilon'[|S_1|^2\,|\Gamma' + \sqrt{\Delta}|^2 - |S_2|^2\,|\Gamma' - \sqrt{\Delta}|^2] \gtreqq 0$$

Table 9

Amplitude reconstruction for reactions of type $\pi p \to K^*\Lambda(0^-1/2^+ \to 1^{-\epsilon}\,1/2^+)$ with polarized target

a) Combined production and joint decay angular distribution, and measurement of the polarization transfer by the method of moments

$$I(\psi, \theta\phi, \theta'\phi') = \frac{1}{2\pi}\sum_{L,L'} C_{K^*}(L)\, C_\Lambda(L')$$

$$X \sum_{M,M'}\left[\overline{t^{LL'}_{MM'}} + P_T\cos\psi\,\overline{{}^z t^{LL'}_{MM'}} + P_T\sin\psi\,\overline{{}^x t^{LL'}_{MM'}} + P_T\,\overline{{}^y t^{LL'}_{MM'}}\right] Y^L_M(\theta,\phi)\, Y^{L'}_{M'}(\theta',\phi')$$

$$C_{K^*}(L)\, C_\Lambda(L')\,(t^{LL'}_{MM'} + P_T\,{}^y t^{LL'}_{MM'}) = \langle Y^L_M(\theta,\phi)\, Y^{L'}_{M'}(\theta',\phi')\rangle$$

$$C_{K^*}(L)\, C_\Lambda(L')\, P_T\,{}^z t^{LL'}_{MM'} = \langle 2\cos\psi\, Y^L_M(\theta,\phi)\, Y^{L'}_{M'}(\theta'\phi')\rangle$$

$$C_{K^*}(L)\, C_\Lambda(L')\, P_T\,{}^x t^{LL'}_{MM'} = \langle 2\sin\psi\, Y^L_M(\theta,\phi)\, Y^{L'}_{M'}(\theta'\phi')\rangle$$

with

$$C_{K^*}(0) = C_\Lambda(0) = 1/\sqrt{4\pi}, \qquad C_{K^*}(2) = -1/\sqrt{2\pi}, \qquad C_\Lambda(1) = \alpha_1/\sqrt{4\pi}$$

(all other $C$ coefficients vanish)

Table 9 — cont'd

---

b) The 48 real observables in transversity quantization

$$\sigma_0,\ t^{20}_{00},\ t^{01}_{00},\ t^{21}_{00},\ \left.\begin{matrix}\mathrm{Re}\\ \mathrm{Im}\end{matrix}\right\}\ t^{20}_{20},\ t^{21}_{20},\ t^{21}_{11},\ t^{21}_{1-1}$$

$$P_R = {}^zt^{00}_{00},\ {}^zt^{20}_{00},\ {}^zt^{01}_{00},\ {}^zt^{21}_{00},\ \left.\begin{matrix}\mathrm{Re}\\ \mathrm{Im}\end{matrix}\right\}\ {}^zt^{20}_{20},\ {}^zt^{21}_{20},\ {}^zt^{21}_{11},\ {}^zt^{21}_{1-1}$$

$$\left.\begin{matrix}\mathrm{Re}\\ \mathrm{Im}\end{matrix}\right\}\ {}^xt^{20}_{10},\ {}^xt^{01}_{01},\ {}^xt^{21}_{10},\ {}^xt^{21}_{01},\ {}^xt^{21}_{01},\ {}^xt^{21}_{2-1}$$

$$\left.\begin{matrix}\mathrm{Re}\\ \mathrm{Im}\end{matrix}\right\}\ {}^yt^{20}_{10},\ {}^yt^{01}_{01},\ {}^yt^{21}_{10},\ {}^yt^{21}_{01},\ {}^yt^{21}_{21},\ {}^yt^{21}_{2-1}$$

($M + M'$ = even for $t^{LL'}_{MM'}$, ${}^zt^{LL'}_{MM'}$ and odd for ${}^xt^{LL'}_{MM'}$, ${}^yt^{LL'}_{MM'}$, by B-symmetry; $t^{LL'}_{-M-M'} = (-)^{M+M'}\, t^{LL'}_{MM'}$)

---

c) The 31 generalized spin rotation parameters and the 16 linear constraints

$$P_0 = \frac{1}{6}\left[1 - \sqrt{10}\, t^{20}_{00} + \sqrt{3}\, t^{01}_{00} - \sqrt{30}\, t^{21}_{00}\right] = \frac{1}{6}\left[P_R - \sqrt{10}\, {}^zt^{20}_{00} + \sqrt{3}\, {}^zt^{01}_{00} - \sqrt{30}\, {}^zt^{21}_{00}\right] = P_0$$

$$P_0' = \frac{1}{6}\left[1 - \sqrt{10}\, t^{20}_{00} - \sqrt{3}\, t^{01}_{00} + \sqrt{30}\, t^{21}_{00}\right] = \frac{-1}{6}\left[P_R - \sqrt{10}\, {}^zt^{20}_{00} - \sqrt{3}\, {}^zt^{01}_{00} + \sqrt{30}\, {}^zt^{21}_{00}\right] = P_0'$$

$$P = \frac{1}{6}\left[1 + \sqrt{\frac{5}{2}}\, t^{20}_{00} - \sqrt{3}\, t^{01}_{00} - \sqrt{\frac{15}{2}}\, t^{21}_{00}\right] = \frac{1}{6}\left[P_R + \sqrt{\frac{5}{2}}\, {}^zt^{20}_{00} - \sqrt{3}\, {}^zt^{01}_{00} - \sqrt{\frac{5}{2}}\, {}^zt^{21}_{00}\right] = P$$

$$P' = \frac{1}{6}\left[1 + \sqrt{\frac{5}{2}}\, t^{20}_{00} + \sqrt{3}\, t^{01}_{00} + \sqrt{\frac{15}{2}}\, t^{21}_{00}\right] = \frac{-1}{6}\left[P_R + \sqrt{\frac{5}{2}}\, {}^zt^{20}_{00} + \sqrt{3}\, {}^zt^{01}_{00} + \sqrt{\frac{5}{2}}\, {}^zt^{21}_{00}\right] = P'$$

$$P_0 + P_0' + 2(P + P') = 1 \qquad\qquad P_R = P_0 - P_0' + 2(P - P')$$

$$Q = \frac{1}{6}\left[\sqrt{15}\, \overline{t^{20}_{20}} - \sqrt{45}\, \overline{t^{21}_{20}}\right] = \frac{1}{6}\left[\sqrt{15}\, \overline{{}^zt^{20}_{20}} - \sqrt{45}\, \overline{{}^zt^{21}_{20}}\right] = Q$$

$$Q' = \frac{1}{6}\left[\sqrt{15}\, \overline{t^{20}_{20}} + \sqrt{45}\, \overline{t^{21}_{20}}\right] = \frac{-1}{6}\left[\sqrt{15}\, \overline{{}^zt^{20}_{20}} + \sqrt{45}\, \overline{{}^zt^{21}_{20}}\right] = Q'$$

$$S_1 = \frac{1}{6}\left[-\sqrt{45}\, \overline{t^{21}_{1-1}}\right]; \qquad \frac{1}{6}\left[-\sqrt{45}\, \overline{{}^zt^{21}_{1-1}}\right] = T_1$$

$$S_2 = \frac{1}{6}\left[\sqrt{45}\, \overline{t^{21}_{11}}\right]; \qquad \frac{-1}{6}\left[\sqrt{45}\, \overline{{}^zt^{21}_{11}}\right] = T_2$$

$$R_0 = \frac{i}{6}\left[-\sqrt{6}\, \overline{{}^yt^{01}_{01}} + \sqrt{60}\, \overline{{}^yt^{21}_{01}}\right] = \frac{1}{6}\left[-\sqrt{6}\, \overline{{}^xt^{01}_{01}} + \sqrt{60}\, \overline{{}^xt^{21}_{01}}\right] = R_0$$

Table 9 — cont'd

$$R_1 = \frac{i}{6}\left[\sqrt{90}\ \overline{{}^{y}t^{21}_{2-1}}\right] = \frac{1}{6}\left[\sqrt{90}\ \overline{{}^{x}t^{21}_{21}}\right] = R_1$$

$$R_2 = \frac{-i}{6}\left[-\sqrt{90}\ \overline{{}^{y}t^{21}_{21}}\right] = \frac{1}{6}\left[-\sqrt{90}\ \overline{{}^{x}t^{21}_{21}}\right] = R_2$$

$$R_3 = \frac{-i}{6}\left[-\sqrt{6}\ \overline{{}^{y}t^{01}_{01}} - \sqrt{15}\ \overline{{}^{y}t^{21}_{01}}\right] = \frac{1}{6}\left[-\sqrt{6}\ \overline{{}^{x}t^{01}_{01}} - \sqrt{15}\ \overline{{}^{x}t^{21}_{01}}\right] = R_3$$

$$A_1 = \frac{i}{6}\left[-\sqrt{\frac{15}{2}}\ \overline{{}^{y}t^{20}_{10}} + \sqrt{\frac{45}{2}}\ \overline{{}^{y}t^{21}_{10}}\right]; \qquad \frac{1}{6}\left[-\sqrt{\frac{15}{2}}\ \overline{{}^{x}t^{20}_{10}} + \sqrt{\frac{45}{2}}\ \overline{{}^{x}t^{21}_{10}}\right] = U_1$$

$$A_2 = \frac{-i}{6}\left[-\sqrt{\frac{15}{2}}\ \overline{{}^{y}t^{20}_{10}} - \sqrt{\frac{45}{2}}\ \overline{{}^{y}t^{21}_{10}}\right]; \qquad \frac{1}{6}\left[-\sqrt{\frac{15}{2}}\ \overline{{}^{x}t^{20}_{10}} - \sqrt{\frac{45}{2}}\ \overline{{}^{x}t^{21}_{10}}\right] = U_2$$

d) Transfer of Polarization

$$\sigma\varrho^{00}_{++} = \sigma_0 P_0(1 + z) \qquad \sigma\varrho^{00}_{--} = \sigma_0 P_0'(1 - z)$$

$$\sigma\varrho^{00e}_{--} = \sigma_0 P(1 + z) \qquad \sigma\varrho^{11e}_{++} = \sigma_0 P'(1 - z)$$

$$\sigma\varrho^{1-1}_{--} = \sigma_0 Q(1 + z) \qquad \sigma\varrho^{1-1}_{++} = \sigma_0 Q'(1 - z)$$

$$\sigma\varrho^{10e}_{-+} = \sigma_0(S_1 + T_1 z) \qquad \sigma\varrho^{10e}_{+-} = \sigma_0(S_2 - T_2 z)$$

$$\sigma\varrho^{00}_{+-} = \sigma_0 R_0(x - iy) \qquad \sigma\varrho^{11e}_{+-} = \sigma_0 R_3(x + iy)$$

$$\sigma\varrho^{1-1}_{-+} = \sigma_0 R_1(x - iy) \qquad \sigma\varrho^{1-1}_{+-} = \sigma_0 R_2(x + iy)$$

$$\sigma\varrho^{10e}_{--} = \sigma_0(U_1 x - iA_1 y) \qquad \sigma\varrho^{10e}_{++} = \sigma_0(U_2 x + iA_2 y)$$

e) Transversity and helicity amplitudes as in Table 8e)

f) Expression of the observables in c) and d) as functions of the transversity amplitudes in Table 8e)

$$2\sigma_0 P_0 = |a|^2 \qquad 2\sigma_0 P_0' = |a'|^2$$

$$2\sigma_0 P = \frac{1}{2}\left(|b|^2 + |c|^2\right) \qquad 2\sigma_0 P' = \frac{1}{2}\left(|b'|^2 + |c'|^2\right.$$

$$2\sigma_0 Q = b\bar{c} \qquad 2\sigma_0 Q' = c'\bar{b}'$$

$$2\sigma_0 S_1 = \frac{1}{2}\left(b\bar{a} - a'\bar{b}'\right) \qquad 2\sigma_0 S_2 = \frac{1}{2}\left(c'\bar{a}' - a\bar{c}\right)$$

$$2\sigma_0 T_1 = \frac{1}{2}\left(b\bar{a} + a'\bar{b}'\right) \qquad 2\sigma_0 T_2 = \frac{1}{2}\left(c'\bar{a}' + a\bar{c}\right)$$

$$2\sigma_0 R_0 = a\bar{a}' \qquad 2\sigma_0 R_3 = \frac{1}{2}\left(c'\bar{b} + b'\bar{c}\right)$$

Table 9 — cont'd

$$2\sigma_0 R_1 = b\bar{b}' \qquad 2\sigma_0 R_2 = c'\bar{c}$$

$$2\sigma_0 U_1 = \frac{1}{2}(b\bar{a}' - a'\bar{c}) \qquad 2\sigma_0 U_2 = \frac{1}{2}(c'\bar{a} - a\bar{b}')$$

$$2\sigma_0 A_1 = \frac{1}{2}(b\bar{a}' + a'\bar{c}) \qquad 2\sigma_0 A_2 = \frac{1}{2}(c'\bar{a} + a\bar{b}')$$

g) Complement to the algebraic reconstruction of amplitudes in Table 8g), for transversally polarized target

$$|b|^2 = 2\sigma_0|S_1 + T_1|^2/P_0 \qquad |b'|^2 = 2\sigma_0|S_1 - T_1|^2/P_0'$$

$$|c|^2 = 2\sigma_0|S_2 - T_2|^2/P_0 \qquad |c'|^2 = 2\sigma_0|S_2 + T_2|^2/P_0'$$

$$\phi_b - \phi_a = \operatorname{Arg}(S_1 + T_1) \qquad \phi_{c'} - \phi_{a'} = \operatorname{Arg}(S_2 + T_2)$$

Table 10

Amplitude reconstruction for reactions of type $\pi p \to \rho N$ ($0^- \, 1/2^+ \to 1^{-\epsilon} \, 1/2^{+\epsilon}$) with polarized target

a) Combined production and decay angular distribution and measurement of the polarization transfer by the method of moments.

$$I(\psi, \theta, \phi) = \frac{1}{2\pi} \sum_L C(L) \sum_M \left[\overline{t^L_M} + P_T \cos\psi \, \overline{{}^z t^L_M} + P_T \sin\psi \, \overline{{}^x t^L_M} + P_L \overline{{}^y t^L_M}\right] Y^L_M(\theta, \phi)$$

$$C(L)\,(t^L_M + P_L{}^y t^L_M) = \langle Y^L_M(\theta, \phi)\rangle$$

$$C(L)\,P_T{}^z t^L_M = \langle 2\cos\psi\, Y^L_M(\theta, \phi)\rangle$$

$$C(L)\,P_T{}^x t^L_M = \langle 2\sin\psi\, Y^L_M(\theta, \phi)\rangle$$

with

$$C(0) = 1/\sqrt{4\pi}, \qquad C(2) = \sqrt{2\pi} \quad \text{(all other } C \text{ coefficients vanish)}$$

b) The 12 real observables in transversity quantization

$$\sigma_0, \quad t_0^2, \quad \operatorname{Re} t_2^2, \quad \operatorname{Im} t_2^2$$

$$P_R = {}^z t_0^0, \quad {}^z t_0^2, \quad \operatorname{Re} {}^z t_2^2, \quad \operatorname{Im} {}^z t_2^2, \quad \operatorname{Re} {}^x t_1^2, \quad \operatorname{Im} {}^x t_1^2, \quad \operatorname{Re} {}^y t_1^2, \quad \operatorname{Im} {}^y t_1^2$$

($M$ = even for $t^L_M$ and ${}^z t^L_M$ $M$ = odd for ${}^x t^L_M$ and ${}^y t^L_M$, by B-symmetry; $t^L_{-M} = (-)^M t^L_M$).

Table 10 — cont'd

c) Generalized spin rotation parameters

$$P_0 = \frac{1}{6}\left[1 + P_R - \sqrt{10}\,(t_0^2 + {}^z t_0^2)\right]$$

$$P_0' = \frac{1}{6}\left[1 - P_R - \sqrt{10}\,(t_0^2 - {}^z t_0^2)\right]$$

$$P = \frac{1}{6}\left[1 + P_R + \sqrt{\frac{5}{2}}\,(t_0^2 + {}^z t_0^2)\right]$$

$$P' = \frac{1}{6}\left[1 - P_R + \sqrt{\frac{5}{2}}\,(t_0^2 - {}^z t_0^2)\right]$$

$$Q = \frac{1}{6}\sqrt{15}\,\left(\overline{t_2^2} + \overline{{}^z t_2^2}\right)$$

$$Q' = \frac{1}{6}\sqrt{15}\,\left(\overline{t_2^2} - \overline{{}^z t_2^2}\right)$$

$$U = \frac{-1}{6}\sqrt{30}\,\overline{{}^x t_1^2}$$

$$A = \frac{-i}{6}\sqrt{30}\,\overline{{}^y t_1^2}$$

d) Transfer of polarization

$$\sigma\varrho_{00} = \sigma_0[(P_0 + P_0') + (P_0 - P_0')\,z]$$

$$\sigma\varrho_{11}^e = \sigma_0[(P + P') + (P - P')\,z]$$

$$\sigma\varrho_{1-1} = \sigma_0[(Q + Q') + (Q - Q')\,z]$$

$$\sigma\varrho_{10}^e = \sigma_0[Ux - iAy]$$

e) Transversity and helicity amplitudes, as in Table 8e)

f) Expression of the observables in c) and d) as functions of the transversity amplitudes

$$2\sigma_0 P_0 = |a|^2 \qquad 2\sigma_0 P_0' = |a'|^2$$

$$2\sigma_0 P = \frac{1}{2}\,(|b|^2 + |c|^2) \qquad 2\sigma_0 P' = \frac{1}{2}\,(|b'|^2 + |c'|^2)$$

$$2\sigma_0 Q = b\bar{c} \qquad 2\sigma_0 Q' = c'\bar{b}'$$

$$2\sigma_0 U = \frac{1}{2}\,(b\bar{a}' - a\bar{b}' + c'\bar{a} - a'\bar{c})$$

$$2\sigma_0 A = \frac{1}{2}\,(b\bar{a}' - a\bar{b}' - c'\bar{a} + a'\bar{c})$$

### 4.1. Reactions of type $\pi p \to K\Lambda$

#### 4.1.1. The Wolfenstein parameters

It is well known that reactions of the type $\pi p \to K\Lambda$ with polarized target and analysis of the final $\Lambda$ polarization are completely described, at fixed energy and momentum transfer, by $\sigma_0$, the unpolarized differential cross section, and $P$, $R$, $A$, the 3 Wolfenstein parameters (cf. ref. [*21*]) which satisfy one quadratic constraint (cf. table 4). Indeed these 4 real numbers supply the whole phenomenological information, namely the differential cross section $\sigma$ and the final polarization components $(X, Y, Z)$ as functions of the initial ones $(x, y, z)$. In the right part of table 4) we show these functions when the polarizations are quantized in s-transversity frames (cf. section 2.1.). The simple inspection of these functions shows that each one of the 4 parameters can be measured twice. That both experimental procedures must supply the same result constitutes a Wolfenstein theorem which will be proven below.

#### 4.2.1. The complex spin rotation parameters

A complete measurement of the reaction $\pi p \to K\Lambda$ involves a measure of the differential cross section and an analysis of the combined angular distribution of the normal to the reaction plane and of the $\Lambda$ decay products, as it was discussed in section 3.3. This angular distribution is given in table 4a) and its moment analysis yields the polarization transfer parameters $t^L_M$, ${}^zt^L_M$, ${}^xt^L_M$, ${}^yt^L_M$ as shown in the same table. There are 8 real, a priori non vanishing, parameters, given in table 4b) (in transversity quantization). They are not independent; they satisfy some linear and quadratic constraints. The method to derive systematically these constraints was given in section 3.3.2. In the present case we first write the B-symmetric matrices $\varrho_0$ and $\varrho_z$, and the B-antisymmetric ones $\varrho_x$ and $\varrho_y$, which appear in the matrix $W$ (eq. 3.9), in the form

$$\varrho_0 = \begin{array}{|c|c|}\hline P_0 & \\ \hline & P_0' \\ \hline\end{array}\,,\quad \varrho_z = \begin{array}{|c|c|}\hline {}^zP_0 & \\ \hline & {}^zP_0' \\ \hline\end{array}\,,\quad \varrho_x = \begin{array}{|c|c|}\hline & {}^xR_0 \\ \hline \overline{{}^xR_0} & \\ \hline\end{array}\,,\quad \varrho_y = \begin{array}{|c|c|}\hline & -i\,{}^yR_0 \\ \hline i\,\overline{{}^yR_0} & \\ \hline\end{array}$$

Then the polarization transfer matrix $W$ reads

$$W = \sigma_0 \quad \begin{array}{cc|c|c|c|c|c} \lambda_p & \lambda_\Lambda & \begin{matrix}+1/2\\+1/2\end{matrix} & \begin{matrix}+1/2\\-1/2\end{matrix} & \begin{matrix}-1/2\\+1/2\end{matrix} & \begin{matrix}-1/2\\-1/2\end{matrix} & \begin{matrix}\lambda_p'\\\lambda_\Lambda'\end{matrix} \\ \hline +1/2 & +1/2 & P_0 + {}^zP_0 & & & {}^xR_0 + {}^yR_0 & \\ \hline +1/2 & +1/2 & & P_0' - {}^zP_0' & \overline{{}^xR_0} - \overline{{}^yR_0} & & \\ \hline -1/2 & +1/2 & & {}^xR_0 - {}^yR_0 & P_0 + {}^zP_0 & & \\ \hline -1/2 & -1/2 & \overline{{}^xR_0} + \overline{{}^yR_0} & & & P_0' + {}^zP_0' & \\ \hline \end{array}$$

It is the direct sum of the external matrix $W_1$ (with $\lambda_p - \lambda_\Lambda =$ even) and the internal one $W_2$ (with $\lambda_p - \lambda_\Lambda =$ odd). The rank of $W$ must be **1** since the other two particles are spinless. This condition imposes either

$$W_2 = 0, \text{ rank } W_1 = 1 \leftrightarrow P_0 = {}^zP_0, \quad P_0' = {}^zP_0', \quad {}^xR_0 = {}^yR_0, \quad |R_0|^2 = P_0P_0',$$

or

$$W_1 = 0, \text{ rank } W_2 = 1 \leftrightarrow P_0 = -{}^zP_0, \quad P_0' = -{}^zP_0', \quad {}^xR_0 = -{}^yR_0, \quad |R_0|^2 = P_0P_0'.$$

In each case there are 4 linear constraints and one quadratic constraint. For our reaction, with relative parity $\eta = \eta_\pi \eta_p \eta_K \eta_\Lambda = +1$, the first alternative is realized, and the corresponding linear constraints are equivalent to the Wolfehstein theorem. The parameters $P_0$ and $P_0'$ are real, the parameter $R_0 = {}^xR_0 = {}^yR_0$ is complex. They are also called "the spin rotation parameters" of the reaction. The definition of complex spin rotation parameters will be useful in the generalizations below. The relation of $P_0$, $P_0'$, $R_0$ with the Wolfenstein's $A$, $P$, $R$ is given in table 4c, where we also show their relations with the polarization transfer multipole parameters and the linear constraints between the latter. Finally, table 4d gives the quadratic constraint in terms of $A$, $P$, $R$ and $P_0$, $P_0'$, $R_0$ and table 4e shows the polarization transfer in terms of the density matrix elements in transversity quantization which are closer to the amplitudes we intend to reconstruct.

### 4.1.3. Reconstruction of amplitudes

Table 4f introduces the terminology for the helicity and transversity amplitudes (c.f. section 2.2.). Their relations with the spin rotation parameters are given in table 4g. Note that with unpolarized target, i.e. by measuring only $\sigma_0$ and $P$ (or $\sigma_0$ and $P_0$) the moduli $a$ and $a'$ of the transversity amplitudes is determined and only their relative phase in ghost.
On the contrary, for the helicity amplitudes, the moduli are not determined with an unpolarized target.

## 4.2. Reactions of type $\pi p \to K\Sigma^*$ ($0^-\ 1/2^+ \to 0^-\ 3/2^+$)

In table 2a, 30 examples of such reactions are listed. For any of them, with unpolarized target but with analysis of the cascade decay of the final baryon, the transversity amplitudes can be measured, up to one ghost phase, by the procedure indicated in table 5. The determination of the ghost phase needs a polarized target and can be performed following the procedure described in table 6.

### 4.2.1. Reactions with unpolarized target

This section is a comment of table 5. Part a) gives the method for measuring the even multipole parameters by a moment analysis of the two body decay of the $\Sigma^*$ (cf. sect. 2.4.1.). Part b) gives the method for measuring all the multipole parameters by a moment analysis of the cascade decay: $\Sigma^* \to \Lambda\pi$, $\Lambda \to p\pi$ (cf. section 2.4.3.). In fact, for $L$ = even one has $L_1 = 0$ and the $\theta_1$, $\phi_1$ or ($\theta_1{}^h$, $\phi_1{}^h$) disappear; then by integrating on $\theta_1$, $\phi_1$ (or $\theta_1{}^h$, $\phi_1{}^h$), the same distribution as in Part a) is recovered, with $C(L, L_1 = 0, J = L) = C^h(L, L_1 = 0, M = 0) = C(L)/\sqrt{4\pi}$. For each quantization frame for the $\Lambda$, the parameters $t_M{}^1$ can be measured twice: with $L_1 = 1$, $J = 0{,}2$ for the canonical frame and with $L_1 = 1$, $M_1 = 0{,}1$ for the helicity frame. The compatibility of these two groups of measurements is a check of the experimental and theoretical assumptions (no biases, spin 3/2 for the $\Sigma^*$ ...) and can be used to reduce the experimental errors.

The a priori non vanishing multipole parameters are listed in part c). Of course all other multipole moments of the cascade angular distribution can be measured too. Their vanishing is a check of parity conservation in the production and in the $\Sigma^* \to \Lambda\pi$ decay. From these values of the multipole parameters, the density matrix elements are easily obtained (cf. eq. (2.8)). In part d) we give explicitly the non-vanishing density matrix elements in transversity quantization. Remark that since the matrix elements are linear combinations of the multipole parameters, they could be obtained directly by the method of moments as mean values, of complicated linear expressions of spherical harmonics. This method can reduce the errors on density matrix elements, and should be applied when amplitude reconstruction is intended. Nevertheless one should perform the checks mentioned above.
The positivity and rank 2 conditions of the $4 \times 4$ density matrix (cf. section 2.3.) impose to its elements the constraints written in part e). Part f) introduces some simple terminology for the transversity and helicity amplitudes which satisfy the B-symmetry conditions (cf. eq. (2.6)). We give also the relation between these amplitudes when the conventions of section 2.1. are used. Finally part g) shows the very simple connection of the transversity amplitudes with the measurable transversity density matrix elements. Remark that the argument of $Q$ and $Q'$ give the relative phases between the amplitudes $a$ and $b$ and between $a'$ and $b'$. But the relative phase between these two groups of amplitudes is ghost. Therefore the moduli of the helicity amplitudes are also ghosts.

### 4.2.2. Reactions with polarized target

The determination of the ghost phase requires an experiment with a polarized target. Table 6 shows the method for measuring all the observables of the reaction and gives the corresponding generalized spin rotation parameters (for comparison, see section 4.1.). Part a) shows the combined production and cascade decay angular distribution for an arbitrary target polarization (cf. section 3.3.4.). Its moment analysis yields the polarization transfer multipole parameters. In part b) we list those which are not a priori vanishing. Of course the other moments of the combined distribution could be measured and should be found compatible with zero.
Part c) shows the linear constraints on the observed transfer multipole parameters (cf. section 3.2.2.) and introduces the linearly independent generalized spin rotation parameters. The real $P$'s and the complex $Q$'s can be measured with an unpolarized target by the left side equations. The complex $R$'s can be measured with a longitudinally polarized target by the left side equations. Besides, all parameters can be measured with a transversally polarized target by the right side equations. Therefore, the experiments with only transverse or only longitudinal target polarization are equivalent: both supply the whole physical information. But the first experiment allows the check of 8 linear constraints; the other 8 constraints can be checked only when both kinds of experiments are performed.
The generalized spin rotation parameters must still satisfy some quadratic constraints and some positivity conditions (cf. section 3.2.2.) given explicitly in part d). These parameters are the coefficients of the polarization transfer from the target polarization to the final particle density matrix as shown in part e).
At last part g) shows the observables in terms of the transversity amplitudes defined in table 5f). The ghost phase between the amplitudes $a$, $b$ and the amplitudes $a'$, $b'$ is contained in the arguments of the parameters $R$, $R'$, $R_1$ and $R_2$. As emphasized above, it can be measured either with transverse or longitudinal target polarization.

### 4.3. Reactions of type $\pi p \to \pi\Delta$ ($0^- 1/2^+ \to 0^- 3/2^{+e}$)

In table 3a, 34 examples of this type of reactions have been listed. Their amplitudes can be completely reconstructed with a transversally polarized target. Table 7 that we now comment gives the recipes for the reconstruction by measuring some generalized spin rotation parameters (cf. section 4.1.).

Part a) shows the combined angular distribution of the normal to the reaction plane and of the $\Delta$ decay products (cf. section 3.3.2.) It allows the measurement of the polarization transfer multipole parameters by a moment analysis as indicated in the same part a). The list of those parameters which are not a priori vanishing is given in part b). Remark that ${}^{\nu}t_1{}^2$ can be measured only with a longitudinally polarized target. Of course all other moments of the combined angular distribution could be measured and should be found compatible with zero, as a check of parity conservation in the reaction and in the $\Delta$ decay.

Part c) introduces the generalized spin rotation parameters which are linear combinations of the transfer multipole parameters and as them could be directly measured by the moment method. The last line of this part c) shows that a longitudinally polarized target provides no new information, only a complex linear constraint can be checked. This part c) uses the same terminology as part c) of table 6 (reaction type $\pi p \to K\Sigma^*$). But for the $\Delta$ we can only measure the even multipole parameters, i.e., expressions of the type $P_1 + P_1'$ given by the left side equations and $P_1 - P_1'$ given by the right side equations; $R_-$ is simply $R - R'$ and can be obtained from the left side and the right side equations, whence the linear constraint.

The spin rotation parameters must satisfy the non linear rank constraints and the positivity conditions written in part d).

Part e) shows the polarization transfer from the initial polarization to the final density matrix.

Part f) introduces some simple terminology for the transversity and helicity amplitudes which must be B-symmetric (cf. eq. (2.6)), and gives their linear relations for the conventions of section 2.1.

Finally part g) exhibits the relations between observables and amplitudes. All moduli and relative phases of the transversity amplitudes are easily obtained, even the relative phase between the amplitudes $a$, $b$ and the amplitudes $a'$, $b'$ can be directly obtained for instance from the expression in brackets.

### 4.4. Reactions of type $\pi p \to K^*\Lambda$ ($0^- 1/2^+ \to 1^{-e}\ 1/2^+$)

Forty examples of such reactions are listed in table 2b. For each of them, with unpolarized target but with analysis of the joint angular distribution of the final decay, the transversity amplitudes can be reconstructed, up to one ghost phase, following the procedure described in table 8. The determination of the ghost phase needs a polarized target and can be performed following the procedure described in table 9.

#### 4.4.1. Experiment with unpolarized target

This section is a comment of table 8. Part a) gives the method for measuring the double multipole parameters by a moment analysis of the joint two body decays of K* and $\Lambda$ (cf. sect. 2.4.2.). Unprimed indices and arguments correspond to K* polarization and decay while the primed ones correspond to those of $\Lambda$. The a priori non vanishing multipole parameters are listed in part b). Of course all other multipole moments of the joint

angular distribution can be measured too. Their vanishing is a check of parity conservation in the production and in the K* decay.
From these values of the double multipole parameters, the joint density matrix is easily obtained (cf. eq. (2.9.)). In part c) we give explicitly the non vanishing elements of the measurable joint density matrix, in transversity quantization. Upper indices refer to the K* transversities and the index e labels elements of the even density matrix; lower indices are twice the $\Lambda$ transversities. Remark that since the density matrix elements are linear expressions of the multipole parameters, they could be obtained directly, by the method of moments, as mean values of similar linear expressions of spherical harmonics. This method reduces the errors on density matrix elements and should be applied when amplitude reconstruction is intended. Nevertheless one should first perform the parity checks mentioned above.
The positivity and rank 2 conditions of the total $6 \times 6$ density matrix (the measured part plus the ghost part) imposes to its measured elements the constraints written in part d). The two equalities are the rank constraints. They are rather cumbersome but they constitute a new check and furthermore they have a diacrital function. Indeed they contain the square root of a complex number $\Delta$ (function of 4th degree in density matrix elements); the constraints decide which of the two possible roots must be chosen, since they will be satisfied for one of the roots and not for the other. This choice eliminates any discrete ambiguity in the reconstruction of the transversity amplitudes. Of course the check of these rank constraints and the possibility of discriminating the two roots require accurate experimental results and hence high statistics.

Part e) introduces some simple terminology for the transversity and helicity amplitudes which satisfy the B-symmetry conditions (cf. eq. (2.6.)). We give also the relations between these amplitudes when the conventions of section 2.1. are used. Finally we give the relation between our transversity amplitudes and those introduced by Byers and Yang, who use a cartesian basis for the spin 1 particle and a transversity quantization axis which violates the Basel convention.

Part f) shows the expressions of the measured observables as functions of the defined transversity amplitudes, and part g) gives the inverse expressions which allow an algebric reconstruction of the amplitudes. Of course the relative phase between the two sets of amplitudes $a$, $b$, $c$ and $a'$, $b'$, $c'$ is ghost and therefore the moduli of the helicity amplitudes cannot be determined with unpolarized target. Remark also that the determination of $|b|$ and $|c|$ from $P$ and $Q$, and similarly for the primed quantities, contains a discrete ambiguity indicated by the sign $\varepsilon$ or $\varepsilon'$. These signs can nevertheless be fixed by the last inequalities of part g), when the choice of the complex square root of $\Delta$ can be done as discussed above.
Another method for amplitude reconstruction is to fit the expressions in part f), imposing for instance that $a$ and $a'$ be real. A mixed method would be to obtain directly by the method of moments the moduli $|a|^2$, $|b \pm c|^2$, $|a'|^2$, $|b' + c'|^2$ and to fit afterwards the relative phase between these sets of amplitudes by using the values of Im $Q$, Im $Q'$, $S_1$ and $S_2$.

### 4.4.2. Experiment with polarized target

The determination of the ghost phase requires a polarized target. Table 9 shows the method for measuring all the observables and introduces the corresponding generalized spin rotation parameters (for comparison see section 4.1.).

Part a) shows the combined production and joint decay angular distribution for an arbitrary target polarization (cf. section 3.3.2.). Its moment analysis yields the polarization transfer joint multipole parameters. In part b) we list those which are not a priori vanishing. They are 48 and are set according to the way they are measured: the first line (12) can be measured with unpolarized target, the two following lines (24) with transverse target polarization, the last line (12) with longitudinal one. Of course, all other (48) parameters can be measured and should be found compatible with zero.

Part c) shows that among those 48 transfer multipole parameters there exist 16 linear constraints (cf. section 3.2.2.) so that, besides the unpolarized differential cross section $\sigma_0$, 31 generalized spin rotation parameters can be defined. The real $P$'s and the complex $Q$'s and $S$'s can be measured with unpolarized target. In addition to these, a transverse polarized target allows the measurement of 16 more parameters the $T$'s, $R$'s and $U$'s, and a longitudinal polarized target yields the $R$'s and $A$'s, i.e. 12 more than with unpolarized target and 4 more than with transversal polarization.
These parameters are the coefficients of the polarization transfer to the final joint density matrix as shown in part d).

Part f) gives the expression of the observables in terms of the transversity amplitudes defined in table 8e). The ghost phase between the two sets of amplitudes $a$, $b$, $c$ and $a'$, $b'$, $c'$ is contained in the $R$'s, $U$'s and $A$'s parameters and can be measured either by longitudinal or by transverse polarization of the target. In this last case, the measurement of the parameters $T_1$ and $T_2$ allows a more direct reconstruction of the transversity amplitudes as given in part g).

### 4.5. Reactions of type $\pi p \to \rho N$ ($0^- \, 1/2^+ \to 1^{-e} \, 1/2^{+e}$)

In Table 3b) 22 examples of this type of reactions are listed. Their transversity amplitudes can be reconstructed with a transversally polarized target up to one ghost phase and some discrete ambiguities. The experiment with longitudinally polarized target supplies two more observables which allow to check two non linear constraints and to eliminate the discrete ambiguities. But the ghost phase could only be obtained from the polarization of the final nucleon[9]).
Table 10, that we now comment, gives the practical recipes for the amplitude reconstruction, by measuring some generalized spin rotation parameters (cf. Section 4.1.).

Part a) shows the combined angular distribution of the normal to the reaction plane and the $\rho$ decay products (cf. Section 3.3.2.). It allows the measurement of the polarization transfer multipole parameters by a moment analysis as indicated in the same part a). The list of these parameters which are not a priori vanishing is given in Part b). Remark that ${}^{v}t_1{}^{2}$ can be measured only with a longitudinally polarized target. Of course all other moments of the combined angular distribution could be measured and should be found compatible with zero, as a check of parity conservation in the reaction and in the $\rho$ decay.

Part c) introduces the generalized spin rotation parameters, which are linear combinations of the transfer multipole parameters, and as them could be directly measured by the moment method. This Part c) uses the same terminology as Part c) of Table 9 (reaction type $\pi p \to K^*\Lambda$). But in the present case we can only measure the polarization of the first particle, i.e. the transfer double multipole parameters with $L' = M' = 0$. They are given by expressions of the type $P_0 + P_0'$ written in the left side equations and of the type $P_0 - P_0'$ written in the right side equations in Table 9c). The parameters $U$ and $A$ correspond to $U_1 + U_2$ and $A_1 - A_2$. Part d) of Table 10 shows the polarization transfer from the target polarization to the final density matrix.

---

[9]) That is a simple application of the Simonius theorem (cf. ref. [9] and Appendix).

The transition amplitudes in this type of reactions are the same as in the reaction type $\pi p \to K^*\Lambda$. Thus we refer to the simple terminology introduced in Table 8e) for the transversity and helicity amplitudes which must be B-symmetric (cf. Eq. (2.6)), and for their linear relations, when the conventions of Section 2.1. are adopted.

Part f) exhibits the relations between the observable spin rotation parameters and the transversity amplitudes. Remark that amplitudes corresponding to opposite polarizations of the final nucleon, i.e. $(ab'c')$ and $(a'bc)$, are never mixed. Therefore the relative phase between those sets of amplitudes is ghost. The situation is equivalent to that of the reaction type $\pi p \to K^*\Lambda$ with unpolarized initial state. From the polarization point of view both reaction types are related by crossing of the baryons. Indeed Table 10f) is obtained from Table 8f) by means of the substitutions $b \leftrightarrow c'$, $c \leftrightarrow b'$, $P \leftrightarrow P'$, $Q \leftrightarrow Q'$, $S_1 \to 1/2(U - A)$, $S_2 \to 1/2(U + A)$. Therefore in our present case, when all the spin rotation parameters are observed (experiment with transverse and longitudinal target polarization), all the moduli and relative phases (up to ghost one) can be unambiguously reconstructed, and two non linear constraints can be checked. For this purpose the expressions in Table 8d) and 8g can be used with the substitutions mentioned above.
When the experiment is only performed with transversally polarized target, the parameter $A$ and the last expression in Table 10f) are ignored, and the diacritical constraints are not available. Then, from the Table 10f), the six moduli can be obtained up to two discrete ambiguities for the moduli of $b$, $c$ and the moduli of $b'$, $c'$. The two relative phases between these couples of amplitudes are unambiguously measurable. The two relative phases between $a$ and $b'$, $c'$, and between $a'$ and $b$, $c$ can be determined from the expression of $U$, up to at most a $2^4$-uple discrete ambiguity.

## 4.6. Reactions of type $\pi p \to K^{**}\Lambda (0^- 1/2^+ \to 2^{+e} 1/2^+)$

Forty examples of this type of reactions can be obtained from Table 2b). Their transversity amplitudes can be reconstructed (up to one ghost phase) with unpolarized target, but with analysis of the joint angular distribution of the final decays. Table 11 gives the practical recipes for this amplitude reconstruction. It is a simple extension of Table 8, which gives the amplitude reconstruction for reaction type $\pi p \to K^*\Lambda$ and has been commented in Section 4.4.1. We refer to these comments, which can be easily applied to Table 11, although we have omitted here the explicit expressions of the 12 non linear constraints and the algebraic expressions for reconstructing the amplitudes. They are very cumbersome and can be obtained from the equations in Table 11c) by elementary algebra. Anyway the simplest method to reconstruct the amplitudes will be a best fit of these expressions as was commented in Section 4.4.1.
The determination of the ghost phase requires a polarized target. The experiments with transverse target polarization and with longitudinal one supplies 72 and 30 new observables including the ghost phase and new constraints. We have not tabulated all these generalized spin rotation parameters. The corresponding Table would be an extension of Table 9. The measurement of only one final polarization is enough to fix the ghost phase (cf. ref. [9]). If only the polarization of $K^{**}$ is measured, Table 12 could be used.

## 4.7. Reactions of type $\pi p \to A_2 N$ $(0^- 1/2^+ \to 2^{+e} 1/2^{+e})$

Twenty two examples of this type of reactions can be obtained from Table 3b). Their transversity amplitudes can be reconstructed (up to one ghost phase) with polarized target and measurement of the $A_2$ polarization. Table 12 gives the practical recipes for this

Table 11

Amplitude reconstruction for reactions of type $\pi p \to K^{**}\Lambda(0^- \ 1/2^+ \to 2^{+\epsilon} \ 1/2^+)$ with unpolarized target

---

a) Joint angular distribution of the K** and Λ decays, and measurement of the double multipole parameters by the method of moments

$$I(\theta, \phi; \theta', \phi') = \sum_{L,L'} C_{K^{**}}(L)\, C_\Lambda(L') \sum_{M,M'} \overline{t^{LL'}_{MM'}}\, Y^L_M(\theta, \phi)\, Y^{L'}_{M'}(\theta', \phi')$$

$$C_{K^{**}}(L)\, C_\Lambda(L')\, t^{LL'}_{MM'} = \langle Y^L_M(\theta, \phi)\, Y^{L'}_{M'}(\theta', \phi')\rangle$$

with: $C_{K^{**}}(0) = C_\Lambda(0) = 1/\sqrt{4\pi}$, $\quad C_\Lambda(1) = \alpha_\Lambda/\sqrt{4\pi}$

$C_{K^{**}}(2) = -\sqrt{5/14\pi}$, $\quad C_{K^{**}}(4) = \sqrt{9/14\pi}$, $\quad$ for $\quad 2 \to 00$ decay

$C_{K^{**}}(2) = -\sqrt{5/56\pi}$, $\quad C_{K^{**}}(4) = -\sqrt{2/7\pi}$, $\quad$ for $\quad 2^+ \to 1^-0^-$ decay

(all other $C$ coefficients vanish)

---

b) The 30 real observables in transversity quantization

$\sigma_0$, $t^{20}_{00}$, $t^{40}_{00}$, $t^{01}_{00}$, $t^{21}_{00}$, $t^{41}_{00}$,

$$\left.\begin{matrix}\text{Re}\\ \text{Im}\end{matrix}\right\} \left\{\begin{matrix} t^{20}_{20},\ t^{40}_{20},\ t^{40}_{40},\ t^{21}_{20},\ t^{41}_{20},\ t^{41}_{40}, \\ t^{21}_{11},\ t^{21}_{1-1},\ t^{41}_{11},\ t^{41}_{1-1},\ t^{41}_{31},\ t^{41}_{3-1}, \end{matrix}\right.$$

($L =$ even, $M + M' =$ even by B-symmetry, $t^{LL'}_{-M-M'} = (-1)^{M+M'}\, \overline{t^{LL'}_{MM'}}$

---

c) Observable elements of this joint density matrix in transversity quantization

$$\left.\begin{matrix} P_0 = \varrho^{00}_{++} \\ P_0' = \varrho^{00}_{--} \end{matrix}\right\} = \frac{1}{10}\left[1 - \sqrt{\frac{50}{7}}\, t^{20}_{00} + \sqrt{\frac{162}{7}}\, t^{40}_{00} \pm \left(\sqrt{3}\, t^{01}_{00} - \sqrt{\frac{150}{7}}\, t^{21}_{00} + \sqrt{\frac{486}{7}}\, t^{41}_{00}\right)\right]$$

$$\left.\begin{matrix} P_1 = \varrho^{11\epsilon}_{--} \\ P_0' = \varrho^{11\epsilon}_{++} \end{matrix}\right\} = \frac{1}{10}\left[1 - \sqrt{\frac{25}{14}}\, t^{20}_{00} - \sqrt{\frac{72}{7}}\, t^{40}_{00} \mp \left(\sqrt{3}\, t^{01}_{00} - \sqrt{\frac{75}{14}}\, t^{21}_{00} - \sqrt{\frac{216}{7}}\, t^{41}_{00}\right)\right]$$

$$\left.\begin{matrix} P_2 = \varrho^{22\epsilon}_{++} \\ P_2' = \varrho^{22\epsilon}_{--} \end{matrix}\right\} = \frac{1}{10}\left[1 + \sqrt{\frac{50}{7}}\, t^{20}_{00} + \sqrt{\frac{9}{14}}\, t^{40}_{00} \pm \left(\sqrt{3}\, t^{01}_{00} + \sqrt{\frac{150}{7}}\, t^{21}_{00} + \sqrt{\frac{27}{14}}\, t^{41}_{00}\right)\right]$$

$$P_0 + P_0' + 2(P_1 + P_1' + P_2 + P_2') = 1$$

$$\left.\begin{matrix} Q_1 = \varrho^{1-1}_{--} \\ Q_1' = \varrho^{1-1}_{++} \end{matrix}\right\} = \frac{1}{10}\left[\sqrt{\frac{75}{7}}\, \overline{t^{20}_{20}} - \sqrt{\frac{180}{7}}\, \overline{t^{40}_{20}} \mp \left(\sqrt{\frac{225}{7}}\, \overline{t^{21}_{20}} - \sqrt{\frac{540}{7}}\, \overline{t^{41}_{20}}\right)\right]$$

$$\left.\begin{matrix} Q_3 = \varrho^{20\epsilon}_{++} \\ Q_3' = \varrho^{20\epsilon}_{--} \end{matrix}\right\} = \frac{1}{10}\left[\sqrt{\frac{50}{7}}\, \overline{t^{20}_{20}} + \sqrt{\frac{135}{14}}\, \overline{t^{40}_{20}} \pm \left(\sqrt{\frac{150}{7}}\, \overline{t^{21}_{20}} + \sqrt{\frac{405}{14}}\, \overline{t^{41}_{20}}\right)\right]$$

$$\left.\begin{matrix} Q_2 = \varrho^{2-2}_{++} \\ Q_2' = \varrho^{2-2}_{--} \end{matrix}\right\} = \frac{1}{10}\left[\sqrt{45}\, \overline{t^{40}_{40}} \pm \sqrt{135}\, \overline{t^{41}_{40}}\right]$$

Table 11 — cont'd

$$S_1 = \varrho^{10e}_{-+} = \frac{1}{10}\left[-\sqrt{75/7}\,\overline{t^{21}_{1-1}} + \sqrt{810/7}\,\overline{t^{41}_{1-1}}\right]$$

$$S_2 = \varrho^{21e}_{-+} = \frac{1}{10}\left[-\sqrt{450/7}\,\overline{t^{21}_{1-1}} - \sqrt{135/7}\,\overline{t^{41}_{1-1}}\right]$$

$$S_3 = \varrho^{10e}_{+-} = \frac{1}{10}\left[\sqrt{75/7}\,\overline{t^{21}_{11}} - \sqrt{810/7}\,\overline{t^{41}_{11}}\right]$$

$$S_4 = \varrho^{21e}_{+-} = \frac{1}{10}\left[\sqrt{450/7}\,\overline{t^{21}_{11}} + \sqrt{135/7}\,\overline{t^{41}_{11}}\right]$$

$$S_5 = \varrho^{2-1e}_{+-} = \frac{1}{10}\left[\sqrt{135}\,\overline{t^{41}_{31}}\right]$$

$$S_6 = \varrho^{2-1e}_{-+} = \frac{1}{10}\left[-\sqrt{135}\,\overline{t^{41}_{3-1}}\right]$$

d) Transversity and helicity amplitudes, and their relations

| $\lambda_\Lambda$ | $\lambda_{K^{**}}$ | $\lambda_p = +\frac{1}{2}$ | $\lambda_p = -\frac{1}{2}$ | | | |
|---|---|---|---|---|---|---|
| $+\frac{1}{2}$ | 2 | | $d$ | | $-D'$ | $D$ |
| | 1 | $c'$ | | | $C$ | $C'$ |
| | 0 | | $a$ | | $-A'$ | $A$ |
| | $-1$ | $b'$ | | | $B$ | $B'$ |
| | $-2$ | | $e$ | | $-E'$ | $E$ |
| $-\frac{1}{2}$ | 2 | $e'$ | | | $E$ | $E'$ |
| | 1 | | $b$ | | $-B'$ | $B$ |
| | 0 | $a'$ | | | $A$ | $A'$ |
| | $-1$ | | $c$ | | $C'$ | $C$ |
| | $-2$ | $d'$ | | | $D$ | $D'$ |
| | | $= T^{\lambda_{K^{**}}}_{2\lambda_\Lambda, 2\lambda_p}$ | | | $= H^{\lambda_{K^{**}}}_{2\lambda_\Lambda, 2\lambda_p}$ | |

$$4(A - iA') = -2a - \sqrt{6}\,(d + e), \qquad 4(A + iA') = -2a' - \sqrt{6}\,(d' + e')$$

$$4(B - iB') = -2(b + c) - 2(d - e), \qquad 4(B + iB') = -2(b' + c') - 2(d' - e')$$

$$4(C - iC') = -2(b + c) + 2(d - e), \qquad 4(C + iC') = -2(b' + c') + 2(d' - e')$$

$$4(D - iD') = -\sqrt{6}\,a - 2(b - c) + (d + e), \qquad 4(D + iD') = -\sqrt{6}\,a' - 2(b' - c') + (d' + e')$$

$$4(E - iE') = -\sqrt{6}\,a + 2(b - c) + (d + e), \qquad 4(E + iE') = -\sqrt{6}\,a' + 2(b' - c') + (d' + c')$$

d') Collinearity constraints

$$\left.\begin{array}{l} A = B = C = D = E = 0 \\ C' = D' = E' = 0 \end{array}\right\} \Leftrightarrow \left\{\begin{array}{l} a = -a',\ d = -d' = \sqrt{\dfrac{3}{2}}\,a + b \\ b = -b' = c = -c',\ e = -e' = \sqrt{\dfrac{3}{2}}\,a - b \end{array}\right.$$

Table 11 — cont'd

---

e) **Expression of the observables in c) and d) as functions of the transversity amplitudes**

$$2\sigma P_0 = |a|^2 \qquad 2\sigma P_0' = |a'|^2$$

$$2\sigma P_1 = \frac{1}{2}(|b|^2 + |c|^2) \qquad 2\sigma P_1' = \frac{1}{2}(|b'|^2 + |c'|^2)$$

$$2\sigma P_2 = \frac{1}{2}(|d|^2 + |e|^2) \qquad 2\sigma P_2' = \frac{1}{2}(|d'|^2 + |e'|^2)$$

$$2\sigma Q_1 = b\bar{c} \qquad 2\sigma Q_1' = c'\bar{b}'$$

$$2\sigma Q_2 = d\bar{e} \qquad 2\sigma Q_2' = e'\bar{d}'$$

$$2\sigma Q_3 = \frac{1}{2}(d\bar{a} + a\bar{e}) \qquad 2\sigma Q_3' = \frac{1}{2}(e'\bar{a}' + a'\bar{d}')$$

$$2\sigma S_1 = \frac{1}{2}(b\bar{a} - a'\bar{b}') \qquad 2\sigma S_2 = \frac{1}{2}(e'\bar{c}' - c\bar{e})$$

$$2\sigma S_3 = \frac{1}{2}(c'\bar{a}' - a\bar{c}) \qquad 2\sigma S_4 = \frac{1}{2}(d\bar{b} - b'\bar{d}')$$

$$2\sigma S_5 = \frac{1}{2}(d\bar{c} - c'\bar{d}') \qquad 2\sigma S_6 = \frac{1}{2}(e'\bar{b}' - b\bar{e})$$

## Table 12

**Amplitude reconstruction for reactions of type $\pi p \to A_2 N(0^- 1/2^+ \to 2^{+\epsilon} 1/2^{+\epsilon})$ with polarized target**

---

a) Combined production and decay angular distribution and measurement of the polarization transfer by the method of moments

$$I(\psi, \theta, \phi) = \frac{1}{2\pi} \sum_L C(L) \sum_M \left[\overline{t^L_M} + P_T\left(\cos\psi\,\overline{{}^z t^L_M} + \sin\psi\,\overline{{}^x t^L_M}\right) + P_L\,\overline{{}^y t^L_M}\right] Y^L_M(\theta, \phi)$$

$$C(L)\,(t^L_M + P_L{}^y t^L_M) = \langle Y^L_M(\theta, \phi)\rangle$$

$$C(L)\,P_T{}^z t^L_M = \langle 2\cos\psi\, Y^L_M(\theta, \phi)\rangle$$

$$C(L)\,P_\Gamma{}^x t^L_M = \langle 2\sin\psi\, Y^L_M(\theta, \phi)\rangle$$

with

$$C(2) = -\sqrt{5/14\pi},\ C(4) = \sqrt{9/14\pi} \quad \text{for} \quad 2 \to 00 \text{ decay}$$

$$C(2) = -\sqrt{5/56\pi},\ C(4) = -\sqrt{2/7\pi} \quad \text{for} \quad 2^+ \to 1^- 0^- \text{ decay}$$

$$C(0) = 1/\sqrt{4\pi} \quad \text{all other } C \text{ coefficients vanish)}$$

Table 12 — cont'd

b) The 30 real observables in transversity quantization

$$\sigma_0,\ t_0^2,\ t_0^4,\ \left.\begin{matrix}\mathrm{Re}\\ \mathrm{Im}\end{matrix}\right\}\ t_2^2,\ t_2^4,\ t_4^4$$

$$P_R = {}^zt_0^0,\ {}^zt_0^2,\ {}^zt_0^4,\ \left.\begin{matrix}\mathrm{Re}\\ \mathrm{Im}\end{matrix}\right\}\ {}^zt_2^2\ {}^zt_2^4,\ {}^zt_4^4,\ {}^xt_1^2,\ {}^xt_1^4,\ {}^xt_3^4$$

$$\left.\begin{matrix}\mathrm{Re}\\ \mathrm{Im}\end{matrix}\right\}\ {}^yt_1^2,\ {}^yt_1^4,\ {}^yt_3^4$$

($M$ = even for $t_M^L$ and ${}^zt_M^L$, $M$ = odd for ${}^xt_M^L$ and ${}^yt_M^L$, by B-symmetry; $t_M^L = (-1)^M\, \overline{t_M^L}$)

c) The 29 generalized spin rotation parameters

$$\left.\begin{matrix}P_0\\ P_0'\end{matrix}\right\} = \frac{1}{10}\left[(1 \pm P_R) - \sqrt{\frac{50}{7}}\,(t_0^2 \pm {}^zt_0^2) + \sqrt{\frac{162}{7}}\,(t_0^4 \pm {}^zt_0^4)\right]$$

$$\left.\begin{matrix}P_1\\ P_1'\end{matrix}\right\} = \frac{1}{10}\left[(1 \pm P_R) - \sqrt{\frac{25}{14}}\,(t_0^2 \pm {}^zt_0^2) - \sqrt{\frac{72}{7}}\,(t_0^4 \pm {}^zt_0^4)\right]$$

$$\left.\begin{matrix}P_2\\ P_2'\end{matrix}\right\} = \frac{1}{10}\left[(1 \pm P_R) + \sqrt{\frac{50}{7}}\,(t_0^2 \pm {}^zt_0^2) + \sqrt{\frac{9}{14}}\,(t_0^4 \pm {}^zt_0^4)\right]$$

$$P_0 + P_0' + 2(P_1 + P_1' + P_2 + P_2') = 1$$

$$P_0 - P_0' + 2(P_1 - P_1' + P_2 - P_2') = P_R$$

$$\left.\begin{matrix}Q_1\\ Q_1'\end{matrix}\right\} = \frac{1}{10}\left[\sqrt{\frac{75}{7}}\,\left(\overline{t_2^2} \pm \overline{{}^zt_2^2}\right) - \sqrt{\frac{180}{7}}\,\left(\overline{t_2^4} \pm \overline{{}^zt_2^4}\right)\right]$$

$$\left.\begin{matrix}Q_3\\ Q_3\end{matrix}\right\} = \frac{1}{10}\left[\sqrt{\frac{50}{7}}\,\left(\overline{t_2^2} \pm \overline{{}^zt_2^2}\right) + \sqrt{\frac{135}{14}}\,\left(\overline{t_2^4} \pm \overline{{}^zt_2^4}\right)\right]$$

$$\left.\begin{matrix}Q_2\\ Q_2'\end{matrix}\right\} = \frac{1}{10}\sqrt{45}\,\left(\overline{t_4^4} + \overline{{}^zt_4^4}\right)$$

$$U_1 = \frac{1}{10}\left[-\sqrt{\frac{50}{7}}\,\overline{{}^xt_1^2} + \sqrt{\frac{540}{7}}\,\overline{{}^xt_1^4}\right]$$

$$U_2 = \frac{1}{10}\left[-\sqrt{\frac{300}{7}}\,\overline{{}^xt_1^2} - \sqrt{\frac{90}{7}}\,\overline{{}^xt_1^4}\right]$$

$$U_3 = \frac{1}{10}\left[-\sqrt{90}\,\overline{{}^xt_3^4}\right]$$

$$A_1 = \frac{i}{10}\left[-\sqrt{\frac{50}{7}}\,\overline{{}^yt_1^2} + \sqrt{\frac{540}{7}}\,\overline{{}^yt_1^4}\right]$$

$$A_2 = \frac{i}{10}\left[-\sqrt{\frac{300}{7}}\,\overline{{}^yt_1^2} - \sqrt{\frac{90}{7}}\,\overline{{}^yt_1^4}\right]$$

$$A_3 = \frac{i}{10}\left[-\sqrt{90}\,\overline{{}^yt_3^4}\right]$$

Table 12 — cont'd

d) Transfer of polarization

$$\sigma\varrho_{00} = \sigma_0[(P_0 + P_0') + (P_0 - P_0')\, z]$$

$$\sigma\varrho^e_{11} = \sigma_0[(P_1 + P_1') + (P_1 - P_1')\, z]$$

$$\sigma\varrho^e_{22} = \sigma_0[(P_2 + P_2') + (P_2 - P_2')\, z]$$

$$\sigma\varrho_{1-1} = \sigma_0[(Q_1 + Q_1') + (Q_1 - Q_1')\, z]$$

$$\sigma\varrho_{2-2} = \sigma_0[(Q_2 + Q_2') + (Q_2 - Q_2')\, z]$$

$$\sigma\varrho_{20} = \sigma_0[(Q_3 + Q_3') + (Q_3 - Q_3')\, z]$$

$$\sigma\varrho_{10} = \sigma_0[U_1 x - iA_1 y]$$

$$\sigma\varrho_{21} = \sigma_0[U_2 x - iA_2 y]$$

$$\sigma\varrho_{2-1} = \sigma_0[U_3 x - iA_3 y]$$

e) Transversity and helicity amplitudes as in Table 11 d)

f) Expression of the observables in c) and d) as functions of the transversity amplitudes

$$2\sigma_0 P_0 = |a|^2 \qquad 2\sigma_0 P_0' = |a'|^2$$

$$2\sigma_0 P_1 = \frac{1}{2}\,(|b|^2 + |c|^2) \qquad 2\sigma_0 P_1' = \frac{1}{2}\,(|b'|^2 + |c'|^2)$$

$$2\sigma_0 P_2 = \frac{1}{2}\,(|d|^2 + |e|^2) \qquad 2\sigma_0 P_2' = \frac{1}{2}\,(|d'|^2 + |e'|^2)$$

$$2\sigma_0 Q_1 = b\bar{c} \qquad 2\sigma_0 Q_1' = c'\bar{b}'$$

$$2\sigma_0 Q_2 = d\bar{e} \qquad 2\sigma_0 Q_2' = e'\bar{d}'$$

$$2\sigma_0 Q_3 = \frac{1}{2}\,(d\bar{a} + a\bar{e}) \qquad 2\sigma_0 Q_3' = \frac{1}{2}\,(a'\bar{d}' + e'\bar{a}')$$

$$2\sigma_0 U_1 = \frac{1}{2}\,(b\bar{a}' - a\bar{b}' + c'\bar{a} - a'\bar{c}) \qquad 2\sigma_0 A_1 = \frac{1}{2}\,(b\bar{a}' - a\bar{b}' - c'\bar{a} + a'\bar{c})$$

$$2\sigma_0 U_2 = \frac{1}{2}\,(d\bar{c}' - c\bar{d}' + e'\bar{b} - b'\bar{e}) \qquad 2\sigma_0 A_2 = \frac{1}{2}\,(d\bar{c}' - c\bar{d}' - e'\bar{b} + b'\bar{e})$$

$$2\sigma_0 U_3 = \frac{1}{2}\,(d\bar{b}' - b\bar{d}' + e'\bar{c} - c'\bar{e}) \qquad 2\sigma_0 A_3 = \frac{1}{2}\,(d\bar{b}' - b\bar{d}' - e'\bar{c} + e'\bar{e})$$

amplitude reconstruction. It is a simple extension of Table 10, which gives the amplitude reconstruction for reaction type $\pi p \to \rho N$ and has been commented in Section 4.5. We refer to these comments, which can be easily applied to Table 12, although we have omitted here the explicit expressions of the non linear constraints and the algebraic expressions for reconstructing the amplitudes. All this can be obtained from Table 12f) by elementary algebra. Anyway the simplest method to reconstruct the amplitudes will be a best fit of these expressions by fixing two arbitrary phases (e.g. $a$ and $a'$ real).
The ghost phase between the sets of amplitudes $(a, b, e, b', c')$, $(a', d', e', b, c)$ could only be measured by analysis of the final nucleon polarization. The situation is equivalent to that of the reaction $\pi p \to K^{**}\Lambda$ with unpolarized initial state. Indeed Table 12f) can be obtained from Table 11e) by the substitutions $b \leftrightarrow c'$, $c \leftrightarrow b'$, $P_1 \leftrightarrow P_1'$, $Q_1 \leftrightarrow Q_1'$, $S_2 \pm S_1 \to U_1, A_1$, $S_5 \pm S_6 \to U_2, A_2$, $S_4 \pm S_3 \to U_3, A_3$.

### 4.8. Reaction of type $\pi p \to K^*\Sigma^*$ $(0^- \, 1/2^+ \to 1^{-e} \, 3/2^+)$

Thirty examples of this type of reactions can be obtained from Table 2 e). Their transversity amplitudes can be reconstructed (up to one ghost phase) with unpolarized target, but with analysis of the joint angular distribution of the $K^*$ decay and the $\Sigma^*$ cascade decay. Table 13 gives the practical recipes for this amplitude reconstruction. It is a simple extension of Table 8, which gives the amplitude reconstruction for reaction type $\pi p \to K^*\Lambda$ and has been commented in Section 4.4.1. We refer to these comments, which can be easily applied to Table 13, although we have omitted here the explicit expressions of the non linear constraints and the algebraic expressions for reconstructing the amplitudes. They can all be obtained from equations in Table 13 f) by elementary algebra. Anyway the simplest method to reconstruct the amplitudes will be a best fit of these expressions as was commented in Section 4.4.1.
The determination of the ghost phase requires a polarized target. The experiments with transverse target polarization and with longitudinal one supplies 122 and 48 new observables, including the ghost phase and new constraints. We have not tabulated all these generalized spin rotation parameters. The corresponding table would be an extension of Table 9.

Table 13

Amplitude reconstruction for reactions of type $\pi p \to K^*\Sigma^*(0^+ \, 1/2^+ \to 1^{+e} \, 3/2^+)$ with unpolarized target

a) Joint angular distribution of the $K^*\Sigma^*$ decays, and measurement of the $L$ and $L'$ even multipole parameters by the method of moments

$$I(\theta, \phi; \theta', \phi') = \sum_{LL'} C_{K^*}(L)\, C_{\Sigma^*}(L') \sum_{M,M'} \overline{t^{LL'}_{MM'}}\, Y^L_M(\theta, \phi)\, Y^{L'}_{M'}(\theta', \phi')$$

$$C_{K^*}(L)\, C_{\Sigma^*}(L')\, t^{LL'}_{MM'} = \langle Y^L_M(\theta, \phi)\, Y^{L'}_{M'}(\theta', \phi') \rangle$$

with

$$C_{K^*}(0) = C_{\Sigma^*}(0) = 1/\sqrt{4\pi}, \qquad C_{K^*}(2) = -1/\sqrt{2\pi}, \qquad C_{\Sigma^*}(2) = -1/\sqrt{4\pi}$$

(all other $C$ coefficients vanish)

Table 13 — cont'd

---

b) Joint angular distribution of the K* decay and Σ* cascade decay and measurement of the $L$ even double multipole parameters by the method of moments

b 1) Canonical frame for the Λ

$$I(\theta, \phi, \theta', \phi'; \theta_1, \phi_1) = \sum_{LL'JL_1} C_{K^*}(L)\, C(L'L_1J) \sum_{MM'N} \langle JL_1\, NM_1 \mid L'M' \rangle$$

$$\overline{t^{LL'}_{MM'}}\, Y^L_M(\theta, \phi)\; Y^J_M(\theta', \phi')\; Y^{L_1}_{M_1}(\theta_1, \phi_1)$$

$$C_{K^*}(L)\, C(L'L_1J)\, t^{LL'}_{MM'} = \sum_{M_1} \langle JL_1NM_1 \mid L'M' \rangle \langle Y^L_M(\theta, \phi)\; Y^J_M(\theta', \phi')\; Y^{L_1}_{M_1}(\theta_1, \phi_1) \rangle$$

b 2) Helicity frame for the Λ

$$I(\theta, \phi, \theta', \phi', \theta_1{}^h, \phi_1{}^h) = \sum_{LL'L_1M_1} C_{K^*}(L)\, C^h(L'L_1M_1) \sum_{MM'} \overline{t^{LL'}_{MM'}}\, Y^L_M(\theta, \phi)$$

$$\sqrt{\frac{2L'+1}{4\pi}}\; \overline{D^{L'}(\phi'\ \theta'\ 0)^{M'}_{M_1}}\; Y^{L_1}_{M_1}(\theta_1{}^h \phi_1{}^h)$$

$$C_{K^*}(L)\, C^h(L'L_1M_1)\, t^{LL'}_{MM'} = \sqrt{\frac{2L'+1}{4\pi}} \langle Y^L_M(\theta, \varphi)\; \overline{D^{L'}(\phi'\theta'0)^{M'}_{M_1}}\; Y^{L'}_{M'}(\theta_1{}^h, \phi_1{}^h) \rangle$$

with $C_{K^*}(L)$ as in a) and $C(L'L_1J)$ and $C^h(L'L_1M_1)$ given in Table 5 b3)

---

c) The 48 observables in transversity quantization

$\sigma,\ t^{20}_{00},\ t^{01}_{00},\ t^{02}_{00},\ t^{03}_{00},\ t^{21}_{00},\ t^{22}_{00},\ t^{23}_{00}$

$$\left.\begin{matrix}\mathrm{Re} \\ \mathrm{Im}\end{matrix}\right\} \begin{matrix} t^{20}_{20},\ t^{02}_{02},\ t^{03}_{02},\ t^{21}_{20},\ t^{22}_{20},\ t^{23}_{20},\ t^{23}_{02},\ t^{22}_{22},\ t^{23}_{22}, \\ t^{22}_{2-2},\ t^{23}_{2-2},\ t^{21}_{11},\ t^{22}_{11},\ t^{23}_{11},\ t^{21}_{1-1},\ t^{22}_{1-1},\ t^{23}_{13}.\ t^{23}_{1-3}, \end{matrix}$$

($L$ = even, $M + M'$ = even by B-symmetry, $t^{LL'}_{-M-M'} = (-1)^{M+M'}\, t^{LL'}_{MM'}$)

d) Observable elements of the joint density matrix in transversity quantization

$$\left.\begin{matrix} P_0 = \varrho^{00}_{11} \\ P_0' = \varrho^{00}_{-1-1} \end{matrix}\right\} = \frac{1}{12}\left[1 - \sqrt{10}\, t^{20}_{00} - \sqrt{5}\, t^{02}_{00} + \sqrt{50}\, t^{22}_{00} \pm \left(\sqrt{\frac{3}{5}}\, t^{01}_{00} - \sqrt{\frac{63}{5}}\, t^{03}_{00} - \sqrt{6}\, t^{21}_{00} + \sqrt{126}\, t^{23}_{00}\right)\right]$$

$$\left.\begin{matrix} P_1 = \varrho^{11e}_{-1-1} \\ P_1' = \varrho^{11e}_{11} \end{matrix}\right\} = \frac{1}{12}\left[1 + \sqrt{\frac{5}{2}}\, t^{20}_{00} - \sqrt{5}\, t^{02}_{00} - \sqrt{\frac{25}{2}}\, t^{22}_{00} \mp \left(\sqrt{\frac{3}{5}}\, t^{01}_{00} - \sqrt{\frac{63}{5}}\, t^{03}_{00}\right.\right.$$

$$\left.\left. + \sqrt{\frac{3}{2}}\, t^{21}_{00} - \sqrt{\frac{63}{2}}\, t^{23}_{00}\right)\right]$$

$$\left.\begin{matrix} P_2 = \varrho^{11e}_{33} \\ P_2' = \varrho^{11e}_{-3-3} \end{matrix}\right\} = \frac{1}{12}\left[1 + \sqrt{\frac{5}{2}}\, t^{20}_{00} + \sqrt{5}\, t^{02}_{00} + \sqrt{\frac{25}{2}}\, t^{22}_{00} \pm \left(\sqrt{\frac{27}{5}}\, t^{01}_{00} + \sqrt{\frac{7}{5}}\, t^{03}_{00}\right.\right.$$

$$\left.\left. + \sqrt{\frac{27}{2}}\, t^{21}_{00} + \sqrt{\frac{7}{2}}\, t^{23}_{00}\right)\right]$$

Table 13 — cont'd

$$\left.\begin{array}{l} P_3 = \varrho^{00}_{-3-3} \\ P_3' = \varrho^{00}_{33} \end{array}\right\} = \frac{1}{12}\left[1 - \sqrt{10}\, t^{20}_{00} + \sqrt{5}\, t^{02}_{00} - \sqrt{50}\, t^{22}_{00} \mp \left(\sqrt{\frac{27}{5}}\, t^{01}_{00} + \sqrt{\frac{7}{5}}\, t^{03}_{00} - \sqrt{54}\, t^{21}_{00} - \sqrt{14}\, t^{23}_{00}\right)\right]$$

$$P_0 + P_0' + 2(P_1 + P_1' + P_2 + P_2') + P_3 + P_3' = 1$$

$$\left.\begin{array}{l} Q_1 = \varrho^{1-1}_{-1-1} \\ Q_1' = \varrho^{1-1}_{11} \end{array}\right\} = \frac{1}{12}\left[\sqrt{15}\, \overline{t^{20}_{20}} - \sqrt{75}\, \overline{t^{22}_{20}} \mp \left(3\overline{t^{21}_{20}} - \sqrt{189}\, \overline{t^{23}_{20}}\right)\right]$$

$$\left.\begin{array}{l} Q_2 = \varrho^{1-1}_{33} \\ Q_2' = \varrho^{1-1}_{-3-3} \end{array}\right\} = \frac{1}{12}\left[\sqrt{15}\, \overline{t^{20}_{20}} + \sqrt{75}\, \overline{t^{22}_{20}} \pm \left(9\overline{t^{21}_{20}} + \sqrt{21}\, \overline{t^{23}_{20}}\right)\right]$$

$$\left.\begin{array}{l} Q_3 = \varrho^{00}_{1-3} \\ Q_3' = \varrho^{00}_{3-1} \end{array}\right\} = \frac{1}{12}\left[\sqrt{10}\, \overline{t^{02}_{02}} - 10\, \overline{t^{22}_{02}} \mp \left(\sqrt{14}\, \overline{t^{03}_{02}} - \sqrt{140}\, \overline{t^{23}_{02}}\right)\right]$$

$$\left.\begin{array}{l} Q_4 = \varrho^{11e}_{3-1} \\ Q_4' = \varrho^{11e}_{1-3} \end{array}\right\} = \frac{1}{12}\left[\sqrt{10}\, \overline{t^{02}_{02}} + 5\, \overline{t^{22}_{02}} \pm \left(\sqrt{14}\, \overline{t^{03}_{02}} + \sqrt{35}\, \overline{t^{23}_{02}}\right)\right]$$

$$\left.\begin{array}{l} Q_5 = \varrho^{1-1}_{-13} \\ Q_5' = \varrho^{1-1}_{-31} \end{array}\right\} = \frac{1}{12}\left[\sqrt{150}\, \overline{t^{22}_{2-2}} \pm \sqrt{210}\, \overline{t^{23}_{2-2}}\right]$$

$$\left.\begin{array}{l} Q_6 = \varrho^{1-1}_{3-1} \\ Q_6' = \varrho^{1-1}_{1-3} \end{array}\right\} = \frac{1}{12}\left[\sqrt{150}\, \overline{t^{22}_{22}} \pm \sqrt{210}\, \overline{t^{23}_{22}}\right]$$

$$S_1 = \varrho^{10e}_{-11} = \frac{-1}{12}\left[6\, \overline{t^{21}_{1-1}} - \sqrt{126}\, \overline{t^{23}_{1-1}}\right]$$

$$\left.\begin{array}{l} S_3 = \varrho^{10e}_{13} \\ S_5 = \varrho^{10e}_{-3-1} \end{array}\right\} = \frac{-1}{12}\left[\sqrt{27}\, \overline{t^{21}_{1-1}} + \sqrt{42}\, \overline{t^{23}_{1-1}} \pm \sqrt{75}\, \overline{t^{22}_{1-1}}\right]$$

$$S_2 = \varrho^{10e}_{1-1} = \frac{1}{12}\left[6\, \overline{t^{21}_{11}} - \sqrt{126}\, \overline{t^{23}_{11}}\right]$$

$$\left.\begin{array}{l} S_4 = \varrho^{10e}_{31} \\ S_6 = \varrho^{10e}_{-1-3} \end{array}\right\} = \frac{1}{12}\left[\sqrt{27}\, \overline{t^{21}_{11}} + \sqrt{42}\, \overline{t^{23}_{11}} \pm \sqrt{75}\, \overline{t^{22}_{11}}\right]$$

$$S_7 = \varrho^{10e}_{-33} = \frac{-1}{12}\left[\sqrt{210}\, \overline{t^{23}_{1-3}}\right]$$

$$S_8 = \varrho^{10e}_{3-3} = \frac{1}{12}\left[\sqrt{210}\, \overline{t^{23}_{13}}\right]$$

Table 13 — cont'd

e) Transversity and helicity amplitudes

| $\lambda_{\Sigma^*}$ | $\lambda_{K^*}$ | $\lambda_p = +\frac{1}{2}$ | $\lambda_p = -\frac{1}{2}$ |
|---|---|---|---|
| $+\frac{3}{2}$ | 1 | $e$ | |
| | 0 | | $d'$ |
| | $-1$ | | |
| $+\frac{1}{2}$ | 1 | | $c'$ |
| | 0 | $a$ | |
| | $-1$ | | $b'$ |
| $-\frac{1}{2}$ | 1 | $b$ | |
| | 0 | | $a'$ |
| | $-1$ | $c$ | |
| $-\frac{3}{2}$ | 1 | | $f'$ |
| | 0 | $d$ | |
| | $-1$ | | $e'$ |

$= T^{\lambda_{K^*}}_{2\lambda_{\Sigma^*},2\lambda_p}$

| | |
|---|---|
| $E$ | $-E'$ |
| $D'$ | $D$ |
| $F$ | $-F'$ |
| $C'$ | $C$ |
| $A$ | $-A'$ |
| $B'$ | $B$ |
| $B$ | $-B'$ |
| $A'$ | $A$ |
| $C$ | $-C'$ |
| $F'$ | $F$ |
| $D$ | $-D'$ |
| $E'$ | $E$ |

$= H^{\lambda_{K^*}}_{2\lambda_{\Sigma^*},2\lambda_p}$

$$4(A - iA') = -\sqrt{2}\,(b + c) - \sqrt{6}\,(e + f) \qquad 4(A + iA') = -\sqrt{2}\,(b' + c') - \sqrt{6}\,(e' + f'),$$

$$4(B - iB') = -\sqrt{2}\,a - (b - c) - \sqrt{6}\,d - \sqrt{3}\,(e - f)$$

$$4(B + iB') = -\sqrt{2}\,a' - (b' - c') - \sqrt{6}\,d' - \sqrt{3}\,(e' - f')$$

$$4(C - iC') = -\sqrt{2}\,a + (b - c) - \sqrt{6}\,d + \sqrt{3}\,(e - f)$$

$$4(C + iC') = -\sqrt{2}\,a' + (b' - c') - \sqrt{6}\,d + \sqrt{3}\,(e' - f')$$

$$4(D - iD') = -\sqrt{6}\,(b + c) + \sqrt{2}\,(e + f) \qquad 4(D + iD') = -\sqrt{6}\,(b' + c') + \sqrt{2}\,(e' + f')$$

$$4(E - iE') = -\sqrt{6}\,a - \sqrt{3}\,(b - c) + \sqrt{2}\,d + (e - f)$$

$$4(E + iE') = -\sqrt{6}\,a' - \sqrt{3}\,(b' - c') + \sqrt{2}\,d' + (e' - f')$$

$$4(F - iF') = -\sqrt{6}\,a + \sqrt{3}\,(b - c) + \sqrt{2}\,d - (e - f)$$

$$4(F + iF') = -\sqrt{6}\,a' + \sqrt{3}\,(b' - c') + \sqrt{2}\,d' - (e' - f')$$

e') Collinearity constraints

$$\left.\begin{array}{l} A' = B' = C' = D' = E' = F' = 0 \\ C = D = E = 0 \end{array}\right\} \Leftrightarrow \left\{\begin{array}{ll} d = d', & a = a' = \sqrt{2}\,(c - b) + \sqrt{3}\,d \\ b = b', & e = e' = \sqrt{3}\,c + \sqrt{2}\,d \\ c = c', & f = f' = \sqrt{3}\,b - \sqrt{2}\,d \end{array}\right.$$

Table 13 — cont'd

f) Expression of the observables in d) as functions of the transversity amplitudes

$$2\sigma P_0 = |a|^2 \qquad 2\sigma P_0' = |a'|^2$$

$$2\sigma P_1 = \frac{1}{2}(|b|^2 + |c|^2) \qquad 2\sigma P_1' = \frac{1}{2}(|b'|^2 + |c'|^2)$$

$$2\sigma P_2 = \frac{1}{2}(|e|^2 + |f|^2) \qquad 2\sigma P_2' = \frac{1}{2}(|e'|^2 + |f'|^2)$$

$$2\sigma P_3 = |d|^2 \qquad 2\sigma P_3' = |d'|^2$$

$$2\sigma Q_1 = b\bar{c} \qquad 2\sigma Q_1' = c'\bar{b}'$$

$$2\sigma Q_2 = e\bar{f} \qquad 2\sigma Q_2' = f'\bar{e}'$$

$$2\sigma Q_3 = a\bar{d} \qquad 2\sigma Q_3' = d'\bar{a}'$$

$$2\sigma Q_4 = \frac{1}{2}(e\bar{b} + f\bar{c}) \qquad 2\sigma Q_4' = \frac{1}{2}(b'\bar{c}' + e'\bar{f}')$$

$$2\sigma Q_5 = b\bar{f} \qquad 2\sigma Q_5' = f'\bar{b}'$$

$$2\sigma Q_6 = e\bar{c} \qquad 2\sigma Q_6' = c'\bar{e}'$$

$$2\sigma S_1 = \frac{1}{2}(b\bar{a} - a'\bar{b}') \qquad 2\sigma S_2 = \frac{1}{2}(c'\bar{a}' - a\bar{c})$$

$$2\sigma S_3 = \frac{1}{2}(c'\bar{d}' - a\bar{f}) \qquad 2\sigma S_5 = \frac{1}{2}(f'\bar{a}' - d\bar{c})$$

$$2\sigma S_4 = \frac{1}{2}(e\bar{a} - d'\bar{b}') \qquad 2\sigma S_6 = \frac{1}{2}(b\bar{d} - a'\bar{e}')$$

$$2\sigma S_7 = \frac{1}{2}(f'\bar{d}' - d\bar{f}) \qquad 2\sigma S_8 = \frac{1}{2}(e\bar{d} - d'\bar{e}')$$

## 4.9. Reactions of type $\pi p \to \rho\Delta$ $(0^- 1/2^+ \to 1^{-e} 3/2^{+e})$

Thirty four examples of this type of reactions can be obtained from Table 3a). Their amplitudes can be completely reconstructed with transversally polarized target and measurement of the joint angular distribution of the final decays. Table 14 gives the practical recipes for this amplitude reconstruction. It is a simple extension of Table 10, which gives the amplitude reconstruction for reaction type $\pi p \to \rho N$ and has been commented in Section 4.5. We refer to these comments, which can be easily applied to Table 14, although we have omitted here the explicit expressions of the non linear constraints and the algebraic expressions for reconstructing the amplitudes. All them can be obtained from Table 14 f) by elementary algebra. Anyway the simplest method to reconstruct the amplitudes will be a best fit of these expressions by fixing one arbitrary phase (e.g. $a$ = real). Remark that the amplitudes are here obtained without any ghost phase, up to the overall one.

Table 14

Amplitude reconstruction for reactions of type $\pi p \to \rho\Delta$ with polarized target

---

a) Combined production and joint decay angular distribution and measurement of the polarization transfer by the method of moments

$$I(\psi; \theta, \phi, \theta', \phi') = \frac{1}{2\pi} \sum_{L,L} C_\rho(L)\, C_\Delta(L')$$

$$\times \sum_{M,M'} \left[\overline{t^{LL'}_{MM'}} + P_T\left(\cos\psi\, \overline{{}^z t^{LL'}_{MM'}} + \sin\psi\, \overline{{}^x t^{LL'}_{MM'}}\right) + P_L \overline{{}^y t^{LL'}_{MM'}}\right] Y^L_M(\theta, \phi)\, Y^{L'}_{M'}(\theta', \phi')$$

$$C_\rho(L)\, C_\Delta(L')\, (t^{LL'}_{MM'} + P_L t^{LL'}_{MM'}) = \langle Y^L_M(\theta, \phi)\, Y^{L'}_{M'}(\theta, \phi')\rangle$$

$$C_\rho(L)\, C_\Delta(L')\, P_T {}^z t^{LL'}_{MM'} = \langle 2 \cos\psi\, Y^L_M(\theta, \phi)\, Y^{L'}_{M'}(\theta', \phi')\rangle$$

$$C_\rho(L)\, C_\Delta(L')\, P_T {}^x t^{LL'}_{MM'} = \langle 2 \sin\psi\, Y^L_M(\theta, \phi)\, Y^{L'}_{M'}(\theta', \phi')\rangle$$

with

$$C_\rho(0) = C_\Delta(0) = 1/\sqrt{4\pi}, \qquad C_\rho(2) = -1/\sqrt{2\pi}, \qquad C_\Delta(2) = -1/\sqrt{4\pi}$$

(all other $C$ coefficients vanish)

---

b) The 72 real observables in transversity quantization

$$\sigma_0,\, t^{20}_{00},\, t^{02}_{00},\, t^{22}_{00},\, \left.\begin{matrix}\mathrm{Re}\\ \mathrm{Im}\end{matrix}\right\}\, t^{20}_{20},\, t^{02}_{02},\, t^{22}_{20},\, t^{22}_{02},\, t^{22}_{22},\, t^{22}_{2-2},\, t^{22}_{11},\, t^{22}_{1-1}$$

$$P_R,\, {}^z t^{20}_{00},\, {}^z t^{02}_{00},\, {}^z t^{22}_{00},\, \left.\begin{matrix}\mathrm{Re}\\ \mathrm{Im}\end{matrix}\right\}\, {}^z t^{20}_{20},\, {}^z t^{02}_{02},\, {}^z t^{22}_{20},\, {}^z t^{22}_{02},\, {}^z t^{22}_{22},\, {}^z t^{22}_{2-2},\, {}^z t^{22}_{11},\, {}^z t^{22}_{1-1}$$

$$\left.\begin{matrix}\mathrm{Re}\\ \mathrm{Im}\end{matrix}\right\}\, {}^x t^{20}_{10},\, {}^x t^{02}_{01},\, {}^x t^{22}_{10},\, {}^x t^{22}_{01},\, {}^x t^{22}_{21},\, {}^x t^{22}_{12},\, {}^x t^{22}_{2-1},\, {}^x t^{22}_{1-2}$$

$$\left.\begin{matrix}\mathrm{Re}\\ \mathrm{Im}\end{matrix}\right\}\, {}^y t^{20}_{10},\, {}^y t^{02}_{01},\, {}^y t^{22}_{10},\, {}^y t^{22}_{01},\, {}^y t^{22}_{21},\, {}^y t^{22}_{12},\, {}^y t^{22}_{2-1},\, {}^y t^{22}_{1-2}$$

---

c) The 63 generalized spin rotation parameters and the 8 linear constraints

$$\left.\begin{matrix}P_0\\ P_0'\end{matrix}\right\} = \frac{1}{12}\left[(1 \pm P_R) - \sqrt{10}\, (t^{20}_{00} \pm {}^z t^{20}_{00}) - \sqrt{5}\, (t^{02}_{00} \pm {}^z t^{02}_{00}) + \sqrt{50}\, (t^{22}_{00} + {}^z t^{22}_{00})\right]$$

$$\left.\begin{matrix}P_1\\ P_1'\end{matrix}\right\} = \frac{1}{12}\left[(1 \pm P_R) + \sqrt{\frac{5}{2}}\, (t^{20}_{00} \pm {}^z t^{20}_{00}) - \sqrt{5}\, (t^{02}_{00} \pm {}^z t^{20}_{00}) - \sqrt{\frac{25}{2}}\, (t^{22}_{00} \pm {}^z t^{22}_{00})\right]$$

$$\left.\begin{matrix}P_2\\ P_2'\end{matrix}\right\} = \frac{1}{12}\left[(1 \pm P_R) + \sqrt{\frac{5}{2}}\, (t^{20}_{00} \pm {}^z t^{20}_{00}) + \sqrt{5}\, (t^{02}_{00} \pm {}^z t^{02}_{00}) + \sqrt{\frac{25}{2}}\, (t^{22}_{00} \pm {}^z t^{22}_{00})\right]$$

$$\left.\begin{matrix}P_3\\ P_3'\end{matrix}\right\} = \frac{1}{12}\left[(1 \pm P_R) - \sqrt{10}\, (t^{20}_{00} \pm {}^z t^{20}_{00}) + \sqrt{5}\, (t^{02}_{00} \pm {}^z t^{02}_{00}) - \sqrt{50}\, (t^{22}_{00} \pm {}^z t^{25}_{00})\right]$$

Table 14 — cont'd

$$P_0 + P_0' + 2(P_1 + P_1' + P_2 + P_2') + P_3 + P_3' = 1$$

$$P_0 - P_0' + 2(P_1 - P_1' + P_2 - P_2') + P_3 - P_3' = P_R$$

$$\left.\begin{matrix} Q_1 \\ Q_1' \end{matrix}\right\} = \frac{1}{12}\left[\sqrt{15}\left(\overline{t^{20}_{20}} \pm \overline{{}^z t^{20}_{20}}\right) - \sqrt{75}\left(\overline{t^{22}_{20}} \pm \overline{{}^z t^{22}_{20}}\right)\right]$$

$$\left.\begin{matrix} Q_2 \\ Q_2' \end{matrix}\right\} = \frac{1}{12}\left[\sqrt{15}\left(\overline{t^{20}_{20}} \pm \overline{{}^z t^{20}_{20}}\right) + \sqrt{75}\left(\overline{t^{22}_{20}} \pm \overline{{}^z t^{22}_{20}}\right)\right]$$

$$\left.\begin{matrix} Q_3 \\ Q_3' \end{matrix}\right\} = \frac{1}{12}\left[\sqrt{10}\left(\overline{t^{02}_{02}} \pm \overline{{}^z t^{02}_{02}}\right) - 10\left(\overline{t^{22}_{02}} \pm \overline{{}^z t^{22}_{02}}\right)\right]$$

$$\left.\begin{matrix} Q_4 \\ Q_4' \end{matrix}\right\} = \frac{1}{12}\left[\sqrt{10}\left(\overline{t^{02}_{02}} \pm \overline{{}^z t^{02}_{02}}\right) + 5\left(\overline{t^{22}_{02}} \pm \overline{{}^z t^{22}_{02}}\right)\right]$$

$$\left.\begin{matrix} Q_5 \\ Q_5' \end{matrix}\right\} = \frac{1}{12}\sqrt{150}\left(\overline{t^{22}_{2-2}} \pm \overline{{}^z t^{22}_{2-2}}\right)$$

$$\left.\begin{matrix} Q_6 \\ Q_6' \end{matrix}\right\} = \frac{1}{12}\sqrt{150}\left(\overline{t^{22}_{22}} \pm \overline{{}^z t^{22}_{2-2}}\right)$$

$$\left.\begin{matrix} T_1 \\ T_1' \end{matrix}\right\} = \frac{-1}{12}\sqrt{75}\left(\overline{t^{22}_{1-1}} \pm \overline{{}^z t^{22}_{1-1}}\right)$$

$$\left.\begin{matrix} T_2 \\ T_2' \end{matrix}\right\} = \frac{1}{12}\sqrt{75}\left(\overline{t^{22}_{11}} \pm \overline{{}^z t^{22}_{11}}\right)$$

c) Continued

$$R_1 = \frac{1}{12}\left[-\sqrt{10}\,\overline{{}^x t^{02}_{01}} + 10\,\overline{{}^x t^{22}_{01}}\right] = \frac{-i}{12}\left[-\sqrt{10}\,\overline{{}^y t^{02}_{01}} + 10\,\overline{{}^y t^{22}_{01}}\right] = R_1$$

$$R_2 = \frac{1}{12}\left[\sqrt{10}\,\overline{{}^x t^{02}_{01}} - 5\,\overline{{}^x t^{22}_{01}}\right] = \frac{i}{12}\left[-\sqrt{10}\,\overline{{}^y t^{02}_{01}} - 5\,\overline{{}^y t^{22}_{01}}\right] = R_2$$

$$R_3 = \frac{1}{12}\left[\sqrt{150}\,\overline{{}^x t^{22}_{2-1}}\right] = \frac{i}{12}\left[\sqrt{150}\,\overline{{}^y t^{22}_{2-1}}\right] = R_3$$

$$R_4 = \frac{1}{12}\left[-\sqrt{150}\,\overline{{}^x t^{22}_{21}}\right] = \frac{i}{12}\left[-\sqrt{150}\,\overline{{}^y t^{22}_{21}}\right] = R_4$$

$$U_1 = \frac{1}{12}\left[-\sqrt{\frac{15}{2}}\,\overline{{}^x t^{20}_{10}} + \sqrt{\frac{75}{2}}\,\overline{{}^x t^{22}_{20}}\right]; \qquad \frac{i}{12}\left[-\sqrt{\frac{15}{2}}\,\overline{{}^y t^{20}_{10}} + \sqrt{\frac{72}{2}}\,\overline{{}^y t^{22}_{10}}\right] = A_1$$

$$U_2 = \frac{1}{12}\left[-\sqrt{\frac{15}{2}}\,\overline{{}^x t^{20}_{10}} - \sqrt{\frac{75}{2}}\,\overline{{}^x t^{22}_{10}}\right]; \qquad \frac{i}{12}\left[-\sqrt{\frac{15}{2}}\,\overline{{}^y t^{20}_{10}} - \sqrt{\frac{75}{2}}\,\overline{{}^y t^{22}_{10}}\right] = A_2$$

$$U_3 = \frac{1}{12}\left[-\sqrt{75}\,\overline{{}^x t^{22}_{1-2}}\right]; \qquad \frac{i}{12}\left[-\sqrt{75}\,\overline{{}^y t^{22}_{1-2}}\right] = A_3$$

$$U_4 = \frac{1}{12}\left[-\sqrt{75}\,\overline{{}^x t^{22}_{12}}\right]; \qquad \frac{i}{12}\left[-\sqrt{75}\,\overline{{}^y t^{22}_{12}}\right] = A_4$$

Table 14 — cont'd

d) Transfer of polarization

$$\sigma\varrho^{00}_{11e} = \sigma_0 \frac{1}{2} [(P_0 + P_0') + (P_0 - P_0')\, z], \qquad \sigma\varrho^{00}_{31e} = \sigma_0 R_1(x + iy),$$

$$\sigma\varrho^{11e}_{11e} = \sigma_0 \frac{1}{2} [(P_1 + P_1') + (P_1 - P_1')\, z], \qquad \sigma\varrho^{11e}_{31e} = \sigma_0 R_2(x - iy),$$

$$\sigma\varrho^{11e}_{33e} = \sigma_0 \frac{1}{2} [(P_2 + P_2') + (P_2 - P_2')\, z], \qquad \sigma\varrho^{1-1}_{13e} = \sigma_0 R_3(x + iy),$$

$$\sigma\varrho^{00}_{33e} = \sigma_0 \frac{1}{2} [(P_3 + P_3') + (P_3 - P_3')\, z], \qquad \sigma\varrho^{1-1}_{31e} = \sigma_0 R_4(x - iy),$$

$$\sigma\varrho^{1-1}_{11e} = \sigma_0 \frac{1}{2} [(Q_1 + Q_1') + (Q_1 - Q_1')\, z], \qquad \sigma\varrho^{10e}_{11e} = \sigma_0(U_1 x - iA_1 y),$$

$$\sigma\varrho^{1-1}_{33e} = \sigma_0 \frac{1}{2} [(Q_2 + Q_2') + (Q_2 - Q_2')\, z], \qquad \sigma\varrho^{10e}_{33e} = \sigma_0(U_2 x - iA_2 y)$$

$$\sigma\varrho^{00}_{3-1e} = \sigma_0 \frac{1}{2} [(Q_3 + Q_3') + (Q_3 - Q_3')\, z], \qquad \sigma\varrho^{10e}_{-13e} = \sigma_0(U_3 x - iA_3 y)$$

$$\sigma\varrho^{11e}_{3-1e} = \sigma_0 \frac{1}{2} [(Q_4 + Q_4') + (Q_4 - Q_4')\, z], \qquad \sigma\varrho^{10e}_{3-1e} = \sigma_0(U_4 x - iA_4 y)$$

$$\sigma\varrho^{1-1}_{-13e} = \sigma_0 \frac{1}{2} [(Q_5 + Q_5') + (Q_5 - Q_5')\, z],$$

$$\sigma\varrho^{1-1}_{3-1e} = \sigma_0 \frac{1}{2} [(Q_6 + Q_6') + (Q_6 - Q_6')\, z],$$

$$\sigma\varrho^{10e}_{13e} = \sigma_0 \frac{1}{2} [(T_1 + T_1') + (T_1 - T_1')\, z],$$

$$\sigma\varrho^{10e}_{31e} = \sigma_0 \frac{1}{2} [(T_2 + T_2') + (T_2 - T_2')\, z],$$

e) Transversity and helicity amplitudes as in Table 13e)

f) Expression of the observables in c) and d) as functions of the transversity amplitudes

$$2\sigma_0 P_0 = |a|^2 \qquad 2\sigma_0 P_0' = |a'|^2$$

$$2\sigma_0 P_1 = \frac{1}{2} (|b|^2 + |c|^2) \qquad 2\sigma_0 P_1' = \frac{1}{2} (|b'|^2 + |c'|^2)$$

$$2\sigma_0 P_1 = \frac{1}{2} (|e|^2 + |f|^2) \qquad 2\sigma_0 P_2' = \frac{1}{2} (|e'|^2 + |f'|^2)$$

$$2\sigma_0 P_2 = |d|^2 \qquad 2\sigma_0 P_3' = |d'|^2$$

Table 14 — cont'd

| | |
|---|---|
| $2\sigma_0 Q_1 = b\bar{c}$ | $2\sigma_0 Q_1' = c'\bar{b}'$ |
| $2\sigma_0 Q_2 = e\bar{f}$ | $2\sigma_0 Q_2' = f'\bar{e}'$ |
| $2\sigma_0 Q_3 = a\bar{d}$ | $2\sigma_0 Q_3' = d'\bar{a}'$ |
| $2\sigma_0 Q_4 = \frac{1}{2}(e\bar{b} + f\bar{e})$ | $2\sigma_0 Q_4' = \frac{1}{2}(b'\bar{e}' + c'\bar{f}')$ |
| $2\sigma_0 Q_5 = b\bar{f}$ | $2\sigma_0 Q_5' = f'\bar{b}'$ |
| $2\sigma_0 Q_6 = e\bar{c}$ | $2\sigma_0 Q_6' = c'\bar{e}'$ |
| $2\sigma_0 T_1 = \frac{1}{2}(d\bar{c} + a\bar{f})$ | $2\sigma_0 T_1' = \frac{1}{2}(f'\bar{a}' + c'\bar{d}')$ |
| $2\sigma_0 T_2 = \frac{1}{2}(e\bar{a} - b\bar{d})$ | $2\sigma_0 T_2' = \frac{1}{2}(a'\bar{e}' + d'\bar{b}')$ |
| $2\sigma_0 R_1 = \frac{1}{2}(d'\bar{a} - a'\bar{d}')$ | $2\sigma_0 R_3 = \frac{1}{2}(c'\bar{f} - f'\bar{c})$ |
| $2\sigma_0 R_2 = \frac{1}{4}(e\bar{c}' - c\bar{e}' + b\bar{f}' - f\bar{b}')$ | $2\sigma_0 R_4 = \frac{1}{2}(e\bar{b}' - b\bar{e}')$ |
| $2\sigma_0 U_1 = \frac{1}{4}(c'\bar{a} - a'\bar{c} + b\bar{a}' - a\bar{b}')$ | $2\sigma_0 U_3 = \frac{1}{4}(f'\bar{a} - a'\bar{f} + b\bar{d}' - d\bar{b}')$ |
| $2\sigma_0 U_2 = \frac{1}{4}(f'\bar{d} - d'\bar{f} + e\bar{d}' - d\bar{e}')$ | $2\sigma_0 U_4 = \frac{1}{4}(c'\bar{d} - d'\bar{c} + e\bar{a}' - a\bar{e}')$ |
| $2\sigma A_1 = \frac{i}{4}(e\bar{c}' - c\bar{e}' - b\bar{f}' + f\bar{b}')$ | $2\sigma_0 A_3 = \frac{i}{4}(f'\bar{a} - a'\bar{f} - b\bar{d}' + d\bar{b}')$ |
| $2\sigma_0 A_2 = \frac{i}{4}(f'\bar{d} - d'\bar{f} - e\bar{d}' + d\bar{e}')$ | $2\sigma_0 A_4 = \frac{i}{4}(c'\bar{d} - d'\bar{c} - e\bar{a}' + a\bar{e}')$ |

## Appendix

### 1. The matrix $W$ of polarization transfer

We consider the linear map $w$ from the initial polarization space of density operators on $\mathscr{H}_e$ to the final polarization space of density operators on $\mathscr{H}_f$:

$$\varrho_e \to w(\varrho_e) \equiv \sigma\varrho_f = T\varrho_e T^\dagger . \tag{A 1}$$

There is a complete mathematical similarity between this polarization transfer and the polarization correlation for a system of two particles. It is therefore very convenient and more elegant to describe polarization transfer by a matrix $W$ analogous to the joint density matrix [*20*]. For an initial polarization density matrix $\varrho_e$ and a final analyser of polarization $A_f$, the transition rate is

$$w(\varrho_e, A_f) = \operatorname{tr} A_f T \varrho_e T^\dagger \tag{A 2}$$

or, writing down the indices (upper = lines, lower = columus),

$$w(\varrho_e, A_f) = \sum_{\substack{\mu\mu' \\ \lambda\lambda'}} (A_f)_{\mu}^{\mu'} \, T_{\lambda}{}^{\mu}(\varrho_e)_{\lambda'}^{\lambda} \, (T^{\dagger})_{\mu'}^{\lambda'} = \sum_{\substack{\mu\mu' \\ \lambda\lambda'}} (\tilde{\varrho}_e)_{\lambda}{}^{\lambda'} \, (A_f)_{\mu}{}^{\mu'} \, (\tilde{T})^{\lambda\mu} \, (\tilde{T}^{\dagger})_{\lambda'\mu'}$$

where the symbol $\sim$ means transposition in the initial space. The last expression can be written

$$w(\varrho_e, A_f) = \operatorname{tr}\,(\tilde{\varrho}_e \otimes A_f)\, W\,, \tag{A3}$$

by using the polarization transfer matrix $W$, whose elements are

$$W_{\lambda'\mu'}^{\lambda\mu} = (\tilde{T})^{\lambda\mu}(\tilde{T}^{\dagger})_{\lambda'\mu'}\,. \tag{A4}$$

This matrix represents an observable, rank one, positive, Hermitian operator acting on the spin space $\mathscr{H}_e {}^*\!\otimes \mathscr{H}_f$. The final state density matrix $\sigma\varrho_f$ is obtained by taking the partial trace in the inital spin space

$$\sigma\varrho_f = \operatorname{tr}_e(\tilde{\varrho}_e \otimes \mathbf{1}_f)\; W\,. \tag{A5}$$

Thus the knowledge of $W$ is equivalent to that of $T$ (or $\tilde{T}$) up to an overall phase. Indeed given any rank one, positive, Hermitian operator, $H$, it is easy to find a vector $|x\rangle$ such that

$$H = |x\rangle\,\langle x|\,. \tag{A6}$$

For any well defined ordering of indices in $H$, let $\lambda^0$ be the first index for which $H_{\lambda^0}^{\lambda^0} \neq 0$. Then a possible vector $|x\rangle$ is defined by components

$$x^{\lambda} = H_{\lambda^0}^{\lambda}/\sqrt{H_{\lambda^0}^{\lambda^0}}\,. \tag{A7}$$

(Remark that $x^{\lambda} = 0$ for $\lambda < \lambda^0$, since $H_{\lambda}{}^{\lambda} = 0$ implies $H_{\lambda'}^{\lambda} = 0 = H_{\lambda}{}^{\lambda'}$ for any $\lambda'$). We call this procedure a "conventional amplitude reconstruction", and denote symbolically

$$|x\rangle = \mathrm{CAR}\,(H)\,. \tag{A7'}$$

Since Eq. (A4) is of the type (A6) with some double indicing, a particular solution of (A4) is the vector $(\tilde{T})^{\lambda\mu}$

$$|\tilde{T}\rangle = \mathrm{CAR}\,(W)\,. \tag{A8}$$

The general solution is obtained by multiplication of this particular one by an arbitrary phase.

Let us study now the structure of the $T$ and $W$ matrices describing two body reactions in which parity is conserved. For this purpose it is convenient to adopt transversity quantizations, for which half of the transition amplitudes vanish (we consider the most frequent case of reactions in which some fermions are present). It is also convenient to introduce a "separation order" for lines and columns of the transition matrix (cf. [*13*]), which segregates the vanishing from the non necessarily vanishing amplitudes.

For one particle with spin $j$ and parity $\eta$, such an ordering classifies the magnetic quantum numbers $\mu$ in two sets $S_e$ and $S_0$:

$$\text{for } j - \mu = \begin{cases} \text{even}, & \mu \in S_e\,, \\ \text{odd}, & \mu \in S_0\,. \end{cases} \tag{A9}$$

Keeping this ordering, the B-symmetry operator (cf. Section 2.2.) for this particle can be written in block form

$$B = \eta e^{-i\pi j} \begin{pmatrix} \mathbf{1}_{j+1,2} & 0 \\ 0 & -\mathbf{1}_{j1/2} \end{pmatrix} \begin{matrix} S_e \\ S_0 \end{matrix} \quad \text{for } j = \text{half odd integer} \tag{A 10a}$$

(columns labelled $S_e$, $S_0$)

$$B = \eta e^{-i\pi j} \begin{pmatrix} \mathbf{1}_{j+1} & 0 \\ 0 & -\mathbf{1}_j \end{pmatrix} \quad \text{for } j = \text{integer}. \tag{A 10b}$$

For a system of two particles with spins $j$, $j'$ and parities $\eta$, $\eta'$ the separation order classifies similarly the couples $(\mu, \mu')$ of magnetic quantum numbers in two sets (cf. [13b]):

$$\text{for } j - \mu + j' - \mu' = \begin{cases} \text{even}, & (\mu, \mu') \in S_e \\ \text{odd}, & (\mu, \mu') \in S_0. \end{cases} \tag{A 11}$$

The corresponding B-symmetry operator reads

$$B = \eta\eta' e^{-i\pi(j+j')} \begin{pmatrix} \mathbf{1}_n & 0 \\ 0 & -\mathbf{1}_n \end{pmatrix} \begin{matrix} S_e \\ S_0 \end{matrix} \tag{A 12}$$

(columns labelled $S_e$, $S_0$)

where $n = 1/2(2j + 1)(2j' + 1)$ was supposed to be integer. For our problem we need two such operators: $B_e$ acting on $\mathscr{H}_e$, the initial spin space of particles 1 and 2, and $B_f$ acting on $\mathscr{H}_f$, the final spin space of particles 3, 4.
Parity conservation in the two body reaction implies

$$B_f T B_e^\dagger = T, \tag{A 13}$$

and imposes to the transition matrix $T$, written in this separation order, one of the two following block structures,

$$T = \begin{pmatrix} T_+ & 0 \\ 0 & T_- \end{pmatrix} \begin{matrix} S_e \\ S_0 \end{matrix}, \quad \text{for } \varepsilon = +1; \qquad T = \begin{pmatrix} 0 & T_+ \\ T_- & 0 \end{pmatrix} \begin{matrix} S_e \\ S_0 \end{matrix}, \quad \text{for } \varepsilon = -1 \tag{A 14}$$

(columns labelled $S_e$, $S_0$)

where

$$\varepsilon = \bar{\eta}_1 \bar{\eta}_2 \eta_3 \eta_4 e^{i\pi(j_1 + j_2 - j_3 - j_4)}. \tag{A 15}$$

By transposition of initial indices, from $T_\pm$ we obtain $\tilde{T}_\pm$, and from $T$ we obtain $T$, which can be written for both values of $\varepsilon$:

$$\tilde{T} = \begin{pmatrix} \frac{1+\varepsilon}{2} \tilde{T}_+ \\ \frac{1-\varepsilon}{2} \tilde{T}_- \\ \frac{1-\varepsilon}{2} \tilde{T}_+ \\ \frac{1+\varepsilon}{2} \tilde{T}_- \end{pmatrix} \begin{matrix} (S_e, S_e) \\ (S_e, S_0) \\ (S_0, S_e) \\ (S_0, S_0). \end{matrix} \tag{A 16}$$

Thus, in this separation order for initial and final indices, the matrix $W$ has the block form

$$W = \tilde{T}\tilde{T}^\dagger = \left(\begin{array}{c|c|c|c} \frac{1+\varepsilon}{2}\tilde{T}_+\tilde{T}_+{}^\dagger & 0 & 0 & \frac{1+\varepsilon}{2}\tilde{T}_+\tilde{T}_-{}^\dagger \\ \hline 0 & \frac{1-\varepsilon}{2}\tilde{T}_+\tilde{T}_-{}^\dagger & \frac{1-\varepsilon}{2}\tilde{T}_+\tilde{T}_+{}^\dagger & 0 \\ \hline 0 & \frac{1-\varepsilon}{2}\tilde{T}_+\tilde{T}_-{}^\dagger & \frac{1-\varepsilon}{2}\tilde{T}_+\tilde{T}_+{}^\dagger & 0 \\ \hline \frac{1+\varepsilon}{2}\tilde{T}_-\tilde{T}_+{}^\dagger & 0 & 0 & \frac{1+\varepsilon}{2}\tilde{T}_-\tilde{T}_-{}^\dagger \end{array}\right) \qquad \text{(A 17)}$$

By further reordering, it could be written as a direct sum of two blocks, one of which necessarily vanishes:

$$W = \frac{1+\varepsilon}{2} W_0 \oplus \frac{1-\varepsilon}{2} W_0, \qquad W_0 = \begin{pmatrix} \tilde{T}_+\tilde{T}_+{}^\dagger & \tilde{T}_+\tilde{T}_-{}^\dagger \\ \tilde{T}_-\tilde{T}_+{}^\dagger & \tilde{T}_-\tilde{T}_-{}^\dagger \end{pmatrix}. \qquad \text{(A 18)}$$

## 2. Number of ghost amplitudes in experiments with unpolarized or polarized spin 1/2 initial particle[10])

We consider now an initial state composed of particle 1 with spin $j_1$, which will be assumed to be unpolarized, and particle 2 with spin $j_2 = 1/2$, whose polarization will be considered. In the main text $j_1 = 0$, and 1 is the beam, 2 the target. This section is more general and can be applied to the case $j_1 = 1/2$ (nucleon scattering), or to the case where 1 is a higher spin nucleus target and 2 a polarized beam. We also assume that the dimension of the final spin space $\mathscr{H}_f$ is not smaller than $2(2j_1 + 1)$, the dimension of the initial one $\mathscr{H}_e$, and that the final density matrix $\sigma\varrho_f$ is completely observed.
According to Eq. (A 12) the B-symmetry operator of the initial state, $B_e$, decomposes in two blocks for a separation order of indices:

$$e^{i\pi(j_1+1/2)}\eta_1\eta_2 B_e = \mathbf{1}_{2j_1+1} \oplus (-\mathbf{1}_{2j_1+1}). \qquad \text{(A 19)}$$

The number $N_U$ of real ghost amplitudes, for unpolarized initial state, is the number of parameters of the set of matrices $U$ which transforms $T$ into $TU$ but leaves the observables unchanged, i.e., $TU(TU)^\dagger = TT^\dagger$. This transformation must preserve the B-symmetric structure of $T$; since furthermore we disregard the overall phase of $T$, the matrices $U$ must satisfy the conditions:

$$UU^\dagger = \mathbf{1}, \qquad B_e U B_e{}^\dagger = U, \qquad \det U = +1. \qquad \text{(A 20)}$$

Because of the structure of $B_e$, Eq. (A 19), these conditions imply that $U$ belongs to the group $S[U(2j_1 + 1) \otimes U(2j_1 + 1)]$. The number $N_U$ of ghost amplitudes is the dimension of this group

$$N_U = 2(2j_1 + 1)^2 - 1. \qquad \text{(A 21)}$$

In the case of an experiment with polarized spin 1/2 initial particle, in addition to Eq. (A 20), $U$ must satisfy the new condition

$$[U, \mathbf{1}_{2j_1+1} \otimes \varrho_2] = 0, \qquad \text{(A 22)}$$

[10]) This section summarizes published and unpublished work of SIMONIUS, cf. [9].

for the density matrix $\varrho_2$ of particle 2. Condition (A 20) and condition (A 22) for arbitrary $\varrho_2$ (we assume that the experiment with longitudinal *and* transverse polarization is performed) are equivalent to

$$U = U_1 \otimes \mathbf{1}_2, \qquad U_1 U_1^\dagger = \mathbf{1}, \qquad \det U_1 = 1, \qquad B_1 U_1 B_1^\dagger = U_1. \tag{A 23}$$

Because of the structure of $B_1$, Eq. (A 10), these conditions imply that $U$ belongs to the group $S[U(j_1 + 1) \otimes U(j_1)]$ for $j_1$ integer, or to the group $S[U(j_1 + 1/2) \otimes U(j_1 + 1/2)]$ for $j_1$ half odd integer. The number $N_P$ of real ghost amplitudes in these cases is given by the dimension of the groups

$$N_P' = \begin{cases} (j_1 + 1)^2 + j_1^2 - 1 = 2j_1(j_1 + 1) & \text{for integer } j_1, \\ 2\left(j_1 + \dfrac{1}{2}\right)^2 - 1 & \text{for half odd integer } j_1. \end{cases} \tag{A 24}$$

Table A 1 gives the value of $N_U$, $N_P$ for the low values of $j_1$. It also gives $N_R = N_U - N_P$, the number of additional amplitudes which can be reached by using a polarized spin 1/2 initial particle.

### 3. The matrices $\varrho_\alpha$ of polarization transfer from spin 1/2 initial particle

We consider further the case of an initial state composed of an unpolarized spin $j_1$ particle and a spin $j_2 = 1/2$ particle whose polarization is described by the density matrix $\varrho_2$. With the usual expansion for $\varrho_2$

$$\varrho_2 = \frac{1}{2}\,(\mathbf{1} + x\tau_x + y\tau_y + z\tau_z), \tag{A 25}$$

the final density matrix can be written

$$\sigma\varrho_f = \sigma_0(\varrho_0 + x\varrho_x + y\varrho_y + z\varrho_z), \tag{A 26}$$

where $\sigma_0$ and $\varrho_0$ are the differential cross section and the final polarization for an unpolarized initial state, and the matrices $\varrho_i (i = x, y, z)$ add the information on the polarization transfer.

The polarization domain of $\varrho_2$ is the Poincaré sphere $x^2 + y^2 + z^2 \leqq 1$; the linear map $w$ transforms this sphere into an ellipsoid centered at $\sigma_0\varrho_0$. The principal axes of this ellipsoid are the new observables which can be measured when the initial state is polarized. They are not arbitrary but must satisfy some conditions, e.g. when rank $\varrho_2 = 1$ (total polarization) rank $\varrho_f \leqq 2j_1 + 1$, and hence the density matrix $\sigma\varrho_f$ (is on the surface of the cone C of positive matrices acting on $\mathscr{H}_f$ (we assume dim $\mathscr{H}_f > 2j_1 + 1$).

For parity conserving two body reactions it is easy to prove that the joint density matrices $\varrho_0$ and $\varrho_z$ are B-symmetric, while $\varrho_x$ and $\varrho_y$ are B-antisymmetric. Indeed the B-symmetry operator for particle 2 [cf. (A 1)],

$$B_2 = -i\eta_2\tau_3, \tag{A 27}$$

decomposes the matrix $\varrho_2$ of (A 24) into

$$\varrho_2 = \varrho_{2+} + \varrho_{2-}, \qquad B_2\varrho_{2\pm}B_2^\dagger = \pm\varrho_{2\pm}, \tag{A 28}$$

with

$$\varrho_{2+} = \frac{1}{2}\,(\mathbf{1} + z\tau_z), \qquad \varrho_{2-} = \frac{1}{2}\,(x\tau_x + y\tau_y). \tag{A 29}$$

Since the density matrix $\varrho_1 = \mathbf{1}_{2j_1+L}/(2j_1 + 1)$ and the transition matrix $T$ are B-symmetric, if we call

$$\sigma\varrho_{f\pm} = T(\varrho_1 \otimes \varrho_{2\pm})\, T^\dagger, \tag{A 30}$$

we obtain

$$B_f \sigma\varrho_{f\pm} B_f^\dagger = \pm\sigma\varrho_{f\pm} \tag{A 31}$$

and

$$\sigma\varrho_{f+} = \sigma_0(\varrho_0 + z\varrho_z), \qquad \sigma\varrho_{f-} = \sigma_0(x\varrho_x + y\varrho_y). \tag{A 32}$$

## 4. Amplitude reconstruction for reactions with spinless beam and spin 1/2 target when the final polarization is completely observed

We suppose for the beam spin $j_1 = 0$. Then the initial state density matrix is that of the target, $\varrho_e = \varrho_z$, with dimension, dim $\mathscr{H}_e = 2$. The most general form for the operator $W$ on $\mathscr{H}_e^* \otimes \mathscr{H}_f$ is

$$W = \sum_\alpha \tilde{\tau}_\alpha \otimes X_\alpha, \tag{A 33}$$

since $\tilde{\tau}_\alpha = \mathbf{1}, \tau_x, \tilde{\tau}_y, \tau_z$ form an orthonormal basis on $\mathscr{H}_e^*$.
Substitution in Eq. (A 5) of the expansion (A 33) of $W$ and (A 25) of $\varrho_e$, use of the identity $\operatorname{tr} \tau_\alpha\tau_\beta = 2\delta_{\alpha\beta}$, and comparison with the definition (A 26) of $\varrho_\alpha$, yield $X_\alpha = \sigma_0\varrho_\alpha$, i.e.,

$$W = \sigma_0 \Sigma_\alpha \tilde{\tau}_\alpha \otimes \varrho_\alpha, \tag{A 34}$$

which is Eq. (3.9) of the main text. Using the ordinary representation of the Pauli matrices, this equation reads

$$W = \sigma_0 \left(\begin{array}{c|c} \varrho_0 + \varrho_z & \varrho_x + i\varrho_y \\ \hline \varrho_x - i\varrho_y & \varrho_0 - \varrho_z \end{array}\right), \tag{A 34'}$$

The separation order is superfluous for the initial spin space, since there are only two indices: $\mu_2 = 1/2 \in S_e$ and $\mu_2 = -1/2 \in S_0$. The matrices $\tilde{T}_\pm$ are then identical to the matrices $T_\pm$. If we introduce the separation order for the final spin space, we may identify the two $W$ expressions (A 32) and (A 34') and we obtain the following block expressions for the polarization transfer matrices:

$$\varrho_0 = \left(\begin{array}{c|c} A & 0 \\ \hline 0 & B \end{array}\right), \quad \varrho_z = \varepsilon \left(\begin{array}{c|c} A & 0 \\ \hline 0 & -B \end{array}\right), \quad \varrho_x = \left(\begin{array}{c|c} & C \\ \hline C^\dagger & \end{array}\right), \quad \varrho_y = \varepsilon \left(\begin{array}{c|c} & -iC \\ \hline iC^\dagger & \end{array}\right) \tag{A 35}$$

with

$$2\sigma_0 A = T_+T_+^\dagger, \qquad 2\sigma_0 B = T_-T_-^\dagger, \qquad 2\sigma_0 C = T_+T_-^\dagger. \tag{A 36}$$

The sign $\varepsilon$ is the function of parities and spins given in (A 15). Remark that $A$, $B$, $C$ are rank one matrices, and $A$, $B$ are Hermitian and positive. The so called "reaction polarization" is

$$P_R \equiv \operatorname{tr} \varrho_z = \varepsilon(\operatorname{tr} A - \operatorname{tr} B). \tag{A 37}$$

Let us study the amplitude reconstruction in the cases in which the final joint polarization can be completely measured, i.e., when one observes $\sigma_0\varrho_0$ in the experiment with unpolarized target, $\sigma_0 \varrho_z$ and $\sigma_0 \varrho_x$ in the experiment with transversally polarized target, and $\sigma_0 \varrho_y$ in the experiment with longitudinally polarized target. In this case, with unpolarized target, all the transversity amplitudes $T_+$ and $T_-$ can be reconstructed up to the overall phase and one ghost phase, as proved by Simonius (cf. Section 2, and Table A1 for $j_1 = 0$). Indeed, by the "conventional amplitude reconstruction" (cf. Section A1), from the observed matrices $\sigma_0 A$ and $\sigma_0 B$ we can obtain

$$|a\rangle = \text{CAR}\,(2\sigma_0 A), \qquad |b\rangle = \text{CAR}\,(2\sigma_0 B), \tag{A38}$$

which are $T_+$ and $T_-$ up to arbitrary phases. The Simonius ghost phase $\varphi$, is the relative phase between $T_+$ and $T_-$. It can only be obtained from $C$, which is observable in experiments with either transverse or longitudinal target polarization. From Eqs. (A36) and (A38), we see that the observed matrix $\sigma_0 C$ must satisfy

$$s\sigma_0 C = |a\rangle \langle b|\, e^{i\varphi}, \tag{A39}$$

and that the amplitude vectors $|a\rangle$ and $e^{-i\varphi}|b\rangle$ can only differ from $T_+$ and $T_-$ by an overall phase, which we disregard.

The number of amplitudes is given by the dimension of $T_\pm$. The number of observables and of their linear and non linear constraints in experiments with different target polarizations are easily obtained from the dimension, Hermiticity and rank properties of the matrices $A$, $B$, $C$. The dimension of the one column matrices $T_\pm$ and the square matrices $A$, $B$, $C$ is

$$n = (2l + 1)\,(j + 1/2), \tag{A40}$$

where $l$ and $j$ are the integer and half odd integer final spins. We recall that a $n \times n$ Hermitian matrix depends on $n^2$ real parameters. If the matrix has rank $k$, these parameters satisfy $(n - k)^2$ constraints of degree $k + 1$. The results are summarized in Table A2. For $l = 0$ equivalent numbers are presented in Table 1. For $l = 0$ and $j = 3/2$, the explicit amplitude reconstruction is presented in Section 4.2. of the main text and in Tables 5—6.

Table A1

Number of ghost amplitudes in reactions with an unpolarized spin $j$ particle and an (unpolarized or polarized) spin 1/2 particle as initial state

| $j$ | 0 | 1/2 | 1 | 3/2 | 2 | integer | half odd integer |
|---|---|---|---|---|---|---|---|
| $N_U$ | 1 | 7 | 17 | 31 | 49 | $8j(j+1)+1$ | $8(j+1/2)^2-1$ |
| $N_P$ | 0 | 1 | 4 | 7 | 12 | $2j(j+1)$ | $2(j+1/2)^2-1$ |
| $N_R$ | 1 | 6 | 13 | 24 | 37 | $6j(j+1)+1$ | $6(j+1/2)^2$ |

The tabulated numbers of real amplitudes are:

$N_U$ = ghost amplitudes for unpolarized initial state
$N_P$ = ghost amplitudes for polarized spin 1/2 initial particle
$N_R = N_U - N_P$ = amplitudes reached by initial polarization.

Note that identity of particles, internal symmetry (isospin charge conjugation) and, for elastic reactions, time reversal may decrease $N_U$.

Table A2

Number of amplitudes, independent observables, and observable constraints in reactions of type $0\ 1/2 \to lj$ ($l$ = integer, $j$ = half odd integer) for different initial polarizations and complete measurement of the final polarization

| $(2l+1)(j+1/2)$ | $A$ | $U_I$ | $U_N$ | $T_I$ | $T_N$ | $T_L$ | $L_I$ | $L_N$ |
|---|---|---|---|---|---|---|---|---|
| 1 | 3 | 2 | 0 | 1 | 1 | 2 | 1 | 1 |
| 2 | 7 | 6 | 2 | 1 | 7 | 8 | 1 | 7 |
| 3 | 11 | 10 | 8 | 1 | 17 | 18 | 1 | 17 |
| 4 | 15 | 14 | 18 | 1 | 31 | 32 | 1 | 31 |
| 5 | 19 | 18 | 32 | 1 | 49 | 50 | 1 | 49 |
| $n$ | $4n-1$ | $4n-2$ | $2(n-1)^2$ | 1 | $2n^2-1$ | $2n^2$ | 1 | $2n^2-1$ |

Terminology:

$A$ = number of real amplitudes (up to the overall phase)
$U$ = number of observables for unpolarized target (or beam)
$T$ = number of new observables reached with transversally polarized target (or beam)
$L$ = idem with longitudinally polarized target (or beam)

The subindices classify these observables into

$I$ = independent observables
$N$ = non linear observable constraints
$L$ = linear observable constraints

(Linear constraints coming from B-symmetry have not been counted, otherwise the total number of observables is multiplied by 2).

## References


[1] M. ABRAMOVICH, A. C. IRVING, A. D. MARTIN, C. MICHAEL, Phys. Letters **39B** (1972) 353. (cf. also the Communication num. 200 to the IInd Aix Conference on Elementary Particles (1973)).

[2] M. AGUILAR-BENITZ, S. U. CHUNG, R. L. EISNER, R. D. FIELD, Phys. Rev. Letters **29** (1972) 749.

[3] R. D. FIELD, R. L. EISNER, S. U. CHUNG, M. AGUILAR-BENITZ, "Transversity Amplitude Analysis of the Reactions $K^-p \to (\omega, \varphi)\ \Lambda$", BNL 17220. (cf. also A. M. COOPER, L. LYONS and A. G. CLARK, Communication num. 179 to the IInd Aix Conference on Elementary Particles (1973)).

[4] R. D. FIELD "Amplitude Analysis of Hypercharge Reactions" BNL 17846 (submitted to Proceedings of Intern. Conf. on $\pi\pi$ Scattering and Associated Topics. Florida State University, March 28–30, 1973). (cf. also the Communication num. 202 to the IInd. Aix Conference on Elementary Particles (1973)).

[5] N. BYERS, C. N. YANG, Phys. Rev. **135** (1964) B 796.

[6] R. W. HUFF, Phys. Rev. **133** (1964) B 1078.

[7] S. M. BERMAN, R. J. OAKES, Phys. Rev. **135** (1964) B 1034.

[8] A. D. MARTIN in "Two Body Collisions" p. 439, Proceedings of the Seventh Rencontre de Moriond, edited by J. TRAN THANH VAN (1972), Paris.

[9] M. SIMONIUS, Phys. Rev. Letters **19** (1967) 279, and "Theory of Polarization Measurements" I and II, Pittsburgh 1969 (unpublished).

[10] M. G. DONCEL, L. MICHEL, P. MINNAERT. "Polarization Density Matrix" PTB Reports No. 35, 37, 44, edited by Laboratoire de Physique Théorique, Université de Bordeaux I, 33170 – Gradignan (France).

1] M. G. DONCEL, L. MICHEL, P. MINNAERT "Matrices Densité de Polarisation", Ecole d'Eté de Physique des Particules – Gif-sur-Yvette, (1970) – edited by R. SALMERON, Ecole Polytechnique, Paris.

[12] P. MINNAERT "The polarization Domain" in "Particle Physics" Les Houches Summer School 1971, edited by C. DE WITT and C. ITZYKSSON, Gordon and Breach (1973), New York.
[13] a) M. G. DONCEL, P. MERY, L. MICHEL, P. MINNAERT, K. C. WALI, Phys. Rev. **D 7** (1973) 815. b) P. MERY, Thèse de 3e cycle, Marseille (1972).
[14] A. KOTANSKI, Acta Phys. Polon. **30** (1966) 629.
[15] G. C. FOX, C. QUIGG, "Production Mechanism of two-to-two scattering processes at inter mediate energies". Annual Review of Nuclear Science, **23**, 219 (1973).
[16] M. R. ROSE, "Elementary Theory of Angular Momentum" Wiley, New York (1957).
[17] M. JACOB, G. C. WICK, Ann. Phys. (N. Y.) **7** (1959) 404.
[18] G. COHEN-TANNOUDJI, A. MOREL, H. NAVELET, Ann. Phys. (N. Y.) **46** (1968) 239.
[19] K. GOTTFRIED, J. D. JACKSON, Nuovo Cimento **34** (1964) 735.
[20] M. G. DONCEL, A. MÉNDEZ, Phys. Letters **41 B** (1972) 83.
[21] L. WOLFENSTEIN, Phys. Rev. **96** (1954) 1654.

# Einführung in die Vielelektronentheorie der Kristalle

Von HUBERTUS STOLZ

1974. XII, 329 Seiten — 44 Abb. — gr. 8° — Leinen M 63,—

Bestell-Nr. 761 503 7 (5890)

Das Buch gibt eine Einführung in die Theorie des Vielteilchensystems der Elektronen in Kristallen, stellt die quantenfeldtheoretischen Methoden zur Behandlung von Vielteilchensystemen allgemein dar und entwickelt den Formalismus der Greenschen Funktionen. Es bringt eine Einführung in die Theorie der Exzitonen und allgemein der elektronischen Elementaranregungen der Kristalle. Die Landau-Theorie der Metallelektronen wird aus den mikroskopischen Grundlagen abgeleitet, und am Beispiel des Pseudopotentials einfacher Metalle wird das Problem der Begründung eines „Einelektronenpotentials" und damit der Blochschen Schrödinger-Gleichung für Kristallelektronen behandelt.

*Bestellungen durch eine Buchhandlung erbeten*

A K A D E M I E - V E R L A G

DDR-108 Berlin, Leipziger Str. 3—4

## INHALT



---

Die „FORTSCHRITTE DER PHYSIK" dienen der Publikation zusammenfassender Berichte über aktuelle Teilgebiete der Physik. Neben Originalberichten in deutscher oder englischer Sprache werden deutsche Übersetzungen wichtiger fremdsprachiger Berichte abgedruckt.

Aus dem Inhalt der folgenden Hefte:

E. B. MANOUKIAN: Vacuum Energy-Density in Quantum Electrodynamics

K. KINOSHITA: Urbaryon Model Approach to Unified Description of Large and Small $P_T$ Spectra

H. MATSUMOTO and H. UMEZAWA: A Rigorous Formulation of the Boson Method in Supercondutivity

G. I. SZÁSZ: Zerfallende Zustände als physikalisch nichtisolierbare Teilsysteme

H. STÖCKEL: Bewegung einer Punktladung in elektromagnetischen Feldern unter Berücksichtigung der Strahlungsdämpfung

Manuskripte werden nach Vereinbarung mit dem Autor von der Schriftleitung entgegengenommen.

Bei Fragen zur Produktsicherheit wenden Sie sich bitte an:
If you have any questions regarding product safety,
please contact:

Walter de Gruyter GmbH
Genthiner Straße 13
10785 Berlin
productsafety@degruyterbrill.com